ENTERPRISE RESOURCE PLANNING (ERP): THE DYNAMICS OF OPERATIONS MANAGEMENT

The cover design was created
by
Mrs. Gili Boker

ENTERPRISE RESOURCE PLANNING (ERP): THE DYNAMICS OF OPERATIONS MANAGEMENT

by

AVRAHAM SHTUB
Industrial Engineering and Management,
Technion, Israel Institute of Technology,
Technion City, Haifa, Israel

Kluwer Academic Publishers
Boston/Dordrecht/London

CONTENTS

Preface

To an increasing extent, corporations are recognizing the strategic role of the operations function. These organizations are discovering that a focus on customer needs is effective only if the operations function is designed and managed to meet those needs. From acquiring raw materials to fabricating parts, to assembling products, to customer delivery, a total systems perspective can enable us, in the ideal, to fashion an operations function like the inner workings of a finely tuned machine (Like clockwork as we used to say in the days before electronic time pieces!).

Life would be uninteresting without change, however, so we can be thankful that operating systems are dynamic in nature. We alter one element and others are affected. We introduce variability at one point and watch the ripple effects over time. These system behaviors can be difficult to grasp and even more difficult to predict.

In addition to understanding the dynamic, integrated nature of systems it is important to understand and to practice the tools supporting the management of these systems. Teaching the concepts of modern information systems and the ability of these systems to enhance competitiveness are an important challenge to any I.E or MBA program.

Modern information systems combine models (in a model base) and data (in a database). These systems are based on an enterprise wide approach to business processes. Enterprise Resource Planning (ERP) software systems provide comprehensive management of financial, manufacturing, sales, distribution and human resources across the enterprise. The ability of ERP systems to support data "drill down" and to eliminate the need to reconcile across functions is designed to enable organizations to compete on the performances of the entire supply chain. To utilize these capabilities managers have to learn how to manage processes using the model base and database of ERP systems. Recognizing this need, some schools are considering the installation of a commercial ERP system and the training of students in using real systems. The amount of time required to learn all the screens and functions of a real ERP system is enormous as these systems are not designed as a teaching tool.

Until now, there has not been an effective mechanism for teaching people to understand the dynamics of operation systems and to practice the usage of ERP systems. This book and the accompanying software fill the void in what we believe to be a very effective manner.

The book has been written with an emphasis on manufacturing firms, but the principles it demonstrates are transferable to other types of operation systems. With this in mind, we have attempted to present service operations issues in the problems at the end of each chapter. The book and the accompanying software have been designed for use in academic and executive programs aimed at teaching students how integrated systems work. Because some basic understanding of operating systems is assumed, we recommend this material for advanced courses in operations management. A course on planning and control systems would probably be the ideal place in business school settings. In an industrial engineering school, the book may give students their first, and perhaps only, introduction to business issues such as market demand and supplier relationships.

We hope that users of this material find it useful and welcome suggestions for its improvement.

ACKNOWLEDGEMENT

I would like to thank the members of the Global Manufacturing Research Group (GMRG) for their comments on the early version of the book. Special thanks to Professor Clay Whybark and Professor Karen Brown for their suggestions and to Beth Lewis for editing and typing the manuscript, and to Tami Lipkin for her administrative help.

1 INTRODUCTION

1.1. Operations management defined

Operations is a term used to define all the activities directly related to the production of goods or services. Operations Management is therefore the function involved in delivering value to the customers. Thus, Operations Management deals with the core business of an organization. Other functions such as marketing, finance, accounting etc. are related to and support the Operations function but are frequently managed separately.

Based on the above definitions, the operations manager manages the process of producing goods and services. Management is a combination of art and science. Shtub et. al. (1994) list seven functions of management: planning, organizing, staffing, directing, motivating , leading and controlling. In this book we focus on three of these functions: planning, directing and controlling the process.

Due to its central role in most organizations, numerous books have been written on Operations Management, and a course in this area is required in most MBA (Master in Business Administration) and in undergraduate I.E. (Industrial Engineering) programs. Most Operations Management books are based on several methodologies of which two are dominant:

*The Operations Research oriented approach - this approach is based on the development of models designed to capture the important aspects of a managerial decision. Tools such as mathematical programming (e.g. linear programming), statistical tools (e.g. regression analysis) and simulation are used to analyze the models and to gain insights to the model and the corresponding problem.

*The case studies oriented approach - this approach is based on the analysis and discussion of specific situations (case studies). Some of the case studies are based on real problems and others are imaginary.

Some books combine the two approaches by presenting models as well as case studies so that the case studies may be analyzed using the models discussed earlier.

Organizations that adopt the above definition of Operations Management are frequently divided along functional lines. The underlying assumption is that the operations people do not have to deal with aspects of marketing, finance and purchasing. Consequently, they need little knowledge about these functional areas while people in these functional areas need to know little about Operations and its management.

In their book Hammer and Champy (1993) suggested the reengineering of business processes as a new approach to inflect changes on organizations. There are few "core processes" in a typical organization:

* The development process - from an idea for a new product or service to a working prototype.
* Preparation of facilities - from a working prototype of a new product or service to the successful completion of design, implementation and testing of the production/assembly or service facility and its supporting systems and resources.
* Sales - from the study of the market and its needs to the reception of a firm customer order.
* Order fulfillment - from a firm customer order to the delivery of the required products or services and payment by customer.
* Service - from customer call for a service to the fixing of the problem and a satisfied customer.

Based on this approach a new role for operations is emerging: managing the order fulfillment process from customer order to the delivery of the goods and services required to achieve satisfaction of the customer. This process is an integrated process as it involves the marketing function- dealing with customers, the purchasing function - dealing with suppliers and subcontractors and the traditional operations - managing the resources required for actual production. This process is dynamic as customer orders may come at any time and for any combination of products and services, customer orders may be modified after reception while the arrival of raw material from suppliers and the availability of resources are subject to uncertainty and changes over time. We define the management of the order fulfillment process as Integrated Production and Order Management- IPOM.

1.2. The need for integrated production and order management

The rapidly changing environment in which the life cycle of products is short and global competition is fierce, forces most organizations to develop adequate policies, tactics and information systems in order to survive. In some industries new product models are introduced every year or two. The technology is ever changing and customers needs are changing along with it. In today's markets, flexibility, quality, cost and time are the four dimensions or cornerstones of competitiveness leading to the survival of an organization and to its success.

Many books and articles discuss time based competition (Blackburn 1991), quality and its management (Deming 1982) and cost. Flexibility, the fourth dimension is achieved by a dynamic approach to Operations and proper integration within and between the different disciplines or functions of the organization.

New managerial approaches are developed to cope with today's markets. In the product development process organizations are able to develop new products, to satisfy changing customer needs in a short time by using Concurrent Engineering

(Nevis and Whitney, 1989), which is an integrated, dynamic approach to new product development.

Concurrent Engineering is based on new product development teams composed of experts from different functions: marketing people who are aware of the customer's needs and expectations. Engineers and designers who know how to translate the customer's needs into specifications, blueprints, assembly instructions etc. Experts in the operations and maintenance of the new product who are focusing on developing a quality, cost-effective product that gives the customer the best value throughout its life cycle. By assigning experts from different functional areas into a team with common goals and objectives, an integrated, dynamic approach to new product development is made possible.

Concurrent Engineering is a team based, integrated, dynamic approach to new product development. A similar approach is needed for the order fulfillment process: an Integrated Production and Order Management approach dealing with the delivery of products or services. This order fulfillment process should yield a fast response to the changing markets, shorter lead times and improved cash flows as a result of efficient use of human resources, money tied in inventories, equipment, information and facilities .

Organizations attempting to cope with the new environment are adopting philosophies such as Just In Time (JIT). This approach (Monden, 1983) is based on low in-process inventories, cross training of workers to achieve workforce flexibility, and high flexibility of machines and equipment that can change from one product to another in a very short time (reduced set up time). Just in time is a process oriented approach – the whole process is integrated by the flow of information through a special set of cards called Kanban as explained in Chapter 6. When implemented successfully, JIT increases the competitiveness of the order fulfillment process, but the approach does not fit all types of organizations and competitive environments and its implementation is difficult and not always successful. Organizations trying to adopt JIT report mixed results - success depends on the kind of industry, the technology used and the competition in the market.

Another approach used to improve flexibility in operations is based on the Pareto Principle, according to which only few of the resources in most manufacturing organizations have a limited capacity and therefore limited flexibility. Management should focus on the scarce resources (bottlenecks) and use their available capacity in the best way to maximize overall performances. This approach known as Drum-Buffer-Rope or DBR was developed by Goldratt and Fox (1986) who extended it later to "The Theory of Constraints".

A third approach is based on dividing the manufacturing organization into a number of focused "cells" each of them specializing in a small number of similar products. This approach known as Group Technology is based on the assumption that it is possible to achieve better performances by managing a small operation. By locating all the facilities required for manufacturing and delivering a family of products in a dedicated cell and by assigning all the people involved in the order fulfillment process of the product family to this "focused " cell, an efficient well managed process is formed. The difficulty with this approach is in the need to physically move machines and other hardware to the same location and to relocate equipment when the family of product changes.

Other approaches to Operations such as agile manufacturing, constant work in process (CONWIP), synchronized manufacturing, and lean manufacturing are discussed in the literature. Each approach represents an attempt to improve competitiveness by integrating and synchronizing the order fulfillment process.

Along with the effort to develop new managerial approaches, information systems that support these new approaches were developed. The early transaction processing systems evolved into Material Requirement Planning (MRP) systems that supports production planning and control. Integration of the MRP logic with modern Data Base Management Systems (DBMS), Decision Support Systems (DSS) and Management Information Systems (MIS) yield the new generation of Enterprise Resource Planning (ERP) systems that manage the data and information requirements of the whole organization. These systems provide the support required for a new, process based approach to management.

ERP systems provide the information and decision support that management was trying to get by other means such as Kanban cards in JIT or the ability to watch the whole process in Group Technology. With ERP systems it is possible to implement Group Technology concepts without the physical relocation of machines, to implement JIT without Kanban cards, or even better to design an order fulfillment process that combines the most appropriate of all the techniques discussed earlier.

Thus the need for a flexible, dynamic, and integrated approach to the management of the order fulfillment process is recognized and information systems that support integration in a dynamic environment are available. However, a proper setting for teaching and testing the different approaches to the management of the process is not yet available. Existing modeling techniques based on mathematical programming are not capable of handling the dynamic nature of the problem, while the case studies approach does not provide an adequate interactive environment to support the teaching of dynamic group decision making and the testing of different managerial philosophies in a dynamic setting.

Integrated Production and Order Management (IPOM) is based on a team approach similar to the new product development team in concurrent engineering. An integrated team that handles the order fulfillment process and is supported by an advanced ERP system is the cornerstone of IPOM.

This book is designed to support the development of teams to implement the Integrated Production and Order Management approach. It is integrated with the Operations Trainer software that provides a dynamic-integrated environment for students and executives to experiment with IPOM and to learn the ERP concepts. The book and the Operations Trainer help to develop the skills needed for a team approach to the management of the order fulfillment process. Our experience with the Operations Trainer and IPOM shows that they promote change, facilitate thinking and support group decision making.

This book and the accompanying software have been designed for use in academic and executive programs, which are aimed at teaching students how integrated systems work. Because some basic understanding of operating systems is assumed, we recommend this material as a supplement for the basic Operations Management course or as a main resource for advanced courses in Operations Management. A course on planning and control systems would probably be the ideal place for IPOM in a business school setting. In an industrial engineering school, this book may give students their first, and perhaps only, introduction to business issues such as market demand and supplier relationships.

1.3. Modeling in operations management

Problem solving and decision making is an important part of the Operation Manager's role. Textbooks and Operations Management courses are frequently organized according to problem types. Thus, in typical Operations Management textbooks there are chapters dealing with inventory related problems, problems of scheduling production, problems in purchasing, etc. These problems are analyzed by an appropriate model and/or discussed using a representative case study.

A model is a simplified presentation of reality. Most real problems are very complex because of sheer size, the number of different factors considered and the dynamic, stochastic (uncertain) nature of the interactions between many of these factors. By making simplifying assumptions it is possible to develop a model of the problem which is simple enough to understand and analyze, and yet provides a good presentation of the real problem.

Many models are mathematical, for example mathematical programming which define an objective function and a set of constraints. Such models are solved by adequate techniques to find the values of the "decision variables" that satisfy the constraints while maximizing (or minimizing) the objective function.

Conceptual models are also common. The organizational structure chart is a frequently used conceptual model describing the relationships between different components of the organization.

When the level of uncertainty is high, statistical models are used to represent the stochastic nature of important factors. Techniques like regression analysis and stochastic dynamic programming are designed to analyze such models.

By analyzing the model, decision-makers are trying to find a good solution to the problem represented by the model. This solution may be useful for the original problem if it is not too sensitive to the simplifying assumptions on which the model is based. Thus, it is important to carry out a sensitivity analysis on the solution obtained to assess its usefulness for the original problem. The relationship between the real problem, the model and the solution are illustrated in Figure 1-1.

The model-based approach to decision making is quite common and many of the software packages for Operations Management are based on this approach. A typical example is STORM (1992) which is available in a student version as well as a commercial version. These models are usually static in the sense that they assume a given value for each input parameter, while in reality many of the input parameters are dynamic - their value is a continuous function of time. Furthermore many of the input parameters are random variables that represent uncertainty in the real problem.

Decision makers adopting the modeling approach have to close the gap between the static models and the dynamic nature of many problems. Implementing the static models periodically whenever a new decision is necessary does this. Thus, it is

required to develop a simulation model of his real system. Models of a job shop and a flow shop are built into the Trainer and by applying a variety of scenarios they serve as examples or case studies from which general insight is gained. The interactive nature of the Operations Trainer simulates a dynamic environment in which the changing values of important parameters are continuously presented to the decision makers and decisions can be made continuously to influence the performances of the system.

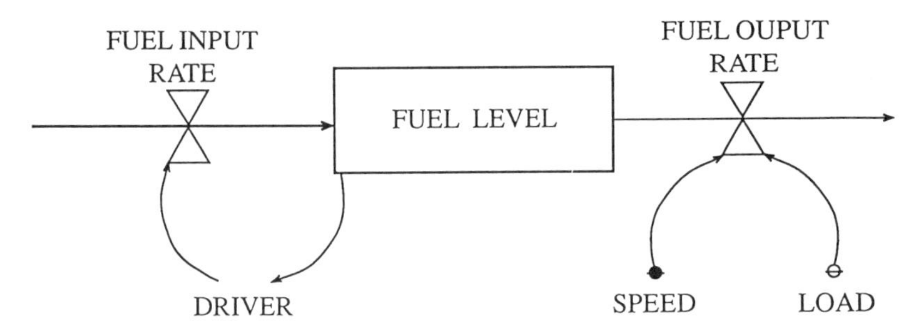

Figure 1.2 Levels and rates in a flow diagram

To capture the dynamic nature of operations, the Trainer is equipped with a simple yet very capable GUI - Graphic User Interface, by which the user can interact and influence the process. Furthermore, the Trainer is integrated in the sense that the different functions such as production, marketing and purchasing share the same database are synchronized and interact over time. Thus, decisions made in the production area influence the rate of raw material consumption as well as the rate of increase in finished goods inventories, which in turn modify the data that influence purchasing and marketing decisions.

By integrating commonly used models for inventory management, production scheduling and sequencing, within an ERP-like system, the Trainer provides a teaching environment in which the modeling approach and the case study approach are integrated in a dynamic sense.

1.5. Integration of simulation with models for decision making

When simulation is used as a decision support tool it should capture three aspects of the real world:

1. The flow of material

2. The flow of information
3. The decision making process

Thus, in a simulation of an inventory system, part of the simulation is devoted to the description of material flow as it enters, stored and leaves the system. Each unit of material is "generated" by a prespecified process (that can be either stochastic i.e. random, or deterministic). The model controls the unit movements in the system. In some models the units of material are "tagged" to facilitate a trace mechanism designed to reveal the history of each unit in the system.

Information regarding the flow of material in the system is collected during the simulation run. The exact time that a tagged unit is generated, moves or leaves the system can be traced. Summary statistics is also available in the form of histograms or frequency tables that summarize the per unit information. The data collected during a simulation run is the basis for understanding and analyzing the inventory system.

The trigger for moves in the system is the decision-making logic built into the simulation model. Such logic, for example, can be based on a simple reorder point model that issues an order for new material whenever the inventory level drops below the "reorder point". By specifying the decision making model, its input parameters and its logic, decisions are built into the simulation model and are automated during simulation runs.

The advantage of this approach is that a simulation run can be performed off-line, i.e. when the decision makers are not present. Thus, it is possible to run large simulation models at night and receive the results the next morning for analysis.

The disadvantage of this approach is that in reality many decisions are based on intuition and experience and it is very difficult (or impossible) to automate these decision making processes. Furthermore, Group Decision-making is a very complex process and so far our knowledge of this process is not sufficient to model it with reasonable accuracy.

To facilitate group decision-making the Operations Trainer is based on an interactive simulation - the processes of material flow and information handling are automated, but decisions are made by the team using the Trainer. A set of decision support models is available for the team, should they decide to use such models. If a "manual" control is selected, each time a decision is required, the information is presented to the decision-makers and the control is shifted to them.

Thus, in the Operations Trainer the computer manages material and information flows while the users make decisions. This approach provides a setting similar to a real life situation where management has the responsibility for making decisions based on information provided by its Management Information Systems (MIS). These decisions, in turn, cause changes in the state of the system.

1.6. Interactive simulation - introduction to the Operations Trainer

The Operations Trainer is a teaching aid designed to facilitate the teaching of Integrated Production and Order Management in the ERP environment. It is based on the following principles:

- A simulation approach - the Trainer simulates a physical system. Models of job shops and flow shops are built into it. The simulation is controlled by a simple user interface and no knowledge of simulation or simulation languages is required.
- A case study approach - the Trainer is based on a simulation of case studies. The details of these case studies are built into the simulation and all the data required for analysis and decision making is easily accessed by the user interface.
- A dynamic approach - the case studies built into the Trainer are dynamic in the sense that the situation changes over time. A random effect is introduced to simulate the uncertainty in the environment, and decisions made by the user cause changes in the state of the system simulated.
- A model based approach - a decision support system is built into the Trainer. This system is based on the ERP concepts of a model base and a data base. The model base contains well known models for forecasting (exponential smoothing), for material management (Material Requirement Planning), for inventory management (a reorder point model), for rough cut capacity planning and for scheduling and sequencing. These models can be consulted at any time or can be used in an "automatic" decision making mode.
- To support decision making further, a data base is built into the Trainer. Data on past performances and data on the current state of the simulated system is readily available to the users. Furthermore, it is possible to use the data as input to the models in the model base to support decision making.
- An integrated approach - Four functions are represented in the Trainer: Finance, Marketing, Production and Purchasing. The four functions are integrated by the simulation, the data base and the model base. Thus, decisions made by any of the four functions affect the state of the simulated system and the data base. Integration within the four functions is an essential feature of ERP systems and the users learn how to work as a team and how to apply group decision-making techniques in this environment.
- User friendliness and GUI - the Trainer is designed as a teaching and training tool. As such, its Graphic User Interface (GUI) is friendly and easy to learn. Although quite complicated scenarios are simulated, and the decision support tools are sophisticated, a typical user can learn how to use the Trainer within a few hours as only the most important features of a real ERP system are available.

By supporting group decision making processes in a dynamic, integrated environment, the Trainer provides the right setting for teaching IPOM within the ERP concept and for developing, evaluating and testing the managerial process adequate for today's competitive environment.

1.7. Overview of the book

This book presents a new approach to the teaching and practicing of Operations Management. The proposed Integrated Production and Order Management (IPOM) approach is process based and ERP supported, unlike the traditional functional based Operations Management approach. The process discussed in this book is the order fulfillment process - from the reception of a customer order to the supply of the right goods on time, the required quantities and at a competitive cost. To teach IPOM a new teaching environment is needed - an environment that supports group decision making in a dynamic setting. An environment that provides models and data relevant to the current state of a dynamic system. The Operations Trainer is designed to provide such an environment.

Thus, this book, along with the Operations Trainer are designed as a complete kit for teaching IPOM. Therefore, the book follows the logic of the Trainer and vice versa.

Each of the following chapters presents a concept of IPOM and the module in the Operations Trainer that supports the teaching of these concepts :

Chapter 2 - **Organizations and Organizational Structures**: The focus of this chapter is on the difference between traditional functional organizational structures and the process based approach of IPOM. The chapter discusses the functional structure and compares it to the project based structures. Each organizational structure is presented, along with its advantages and disadvantages. The matrix structure is also presented as a compromise between the functional structure and the project structure. Concurrent Engineering is discussed as a process oriented approach to new product development. The idea of a multidisciplinary team responsible for a complete process and supported by a common information system which is demonstrated by Concurrent Engineering is later implemented in the order fulfillment process.

Two common layouts or physical structures for manufacturing facilities are discussed next - the job shop and the flow shop. The relationship between the physical layout and the organizational structure is illustrated by introducing the concepts of Group Technology and manufacturing cells.

The role of Operations Management in the functional organization and its interaction with other functions is discussed in order to explain the need for a different approach - the process based approach. The basics of the process based, integrated approach are presented and illustrated by discussing the functions and scenarios built into the Operations Trainer.

Chapter 3 - **Information and its Use**: This chapter focuses on the relationship between information, decision making and the ERP concept. The difference between recurrent decisions that can be automated by adopting a proper policy and one-time decisions that need special management attention and ad-hoc actions is discussed. The message is that a well-designed process is accompanied by a system that collects relevant data and presents it as useful information that is the basis for policies and actions.

The terms data and information are discussed first, data sources, data processing and data storage and retrieval are defined and explained.

The concept of MIS - Management Information System is presented next with an emphasis on the ERP approach - separation of the data base from the model base. Data sharing among different functions in a functional organization is also discussed.

The accounting system is used as an example of a typical MIS and the concepts of data collection and processing are illustrated. Using this example the problem of setting goals and performance measures in the functional organization is discussed. Alternatives to the traditional accounting system are presented to show how the same raw data can be collected, processed, and analyzed by different models. The quality of information and relevant performance measures are also discussed.

Uncertainty and the difficulty of forecasting future data needed for decision making are the subject of the next section. The idea of time series analysis and decomposition of a time series into its major components are explained along with techniques such as moving averages, weighted moving averages and exponential smoothing used for forecast. The trade-off between the cost of data and its accuracy is discussed and common measures for forecasting errors such as the BIAS and MAD are presented.

The last section of this chapter presents the accounting system built into the Operations Trainer. The analysis of product cost and contract cost as performed by the Trainer is explained and illustrated by examples.

Chapter 4 - **Marketing Considerations**: The relationship between Operations and Marketing in a functional organization is the link between the customer and the delivery system. In the integrated order fulfillment process, from a strategic viewpoint, the question facing the order fulfillment team is how to reduce cost and lead time while increasing the service level. Policies such as "make to stock", "make to order" and "assemble to order" are discussed and their effect on cost and lead time are explained in this chapter.

The interface between marketing and operations is the MPS - Master Production Schedule, the concept of MPS and its management are explained and discussed. Competition and its pressure on Marketing and Operations to cut lead times (time based competition) to increase quality (Total Quality Management), to cut cost (cost based competition) and to increase flexibility to changing customer needs and market conditions are discussed next. The relationship between the Operations strategy and the achievement of a competitive advantage are illustrated by the Just In Time (JIT) and the Drum Buffer Rope (Goldratt and Fox, 1986.) approaches.

The last part of this chapter presents the marketing aspects in the Operations Trainer. Product marketing policies related to cost and lead times, product mix considerations and the analysis of large customer orders is discussed. Market segmentation in the context of small customers ordering few units at a given catalog price, vs. major customers purchasing large quantities over a long period of time by long term contracts is explained.

Chapter 5 - **Purchasing and Inventory Management**: In this chapter purchasing policies are discussed. Common models for inventory management are presented along with their underlying assumptions, and a discussion of their advantages and disadvantages. The link between production planning and control and purchasing is explained. The concept of a value chain that links suppliers and manufacturers to the market by means of a logistic system is the backbone of this chapter. Focusing on Value Added activities along the supply chain, and understanding the significance of reliable sources and shorter lead times are two of the chapter's objectives. By introducing the concepts of inventory costs: Capital related costs, costs of inventory management operations and cost of risks related to inventories, the reader gets a feeling for the economic performance measures used for inventory management. Other performance measures such as inventory turns and service level are also discussed.

The question of outsourcing - make or buy is presented along with a discussion on break-even analysis.

Dealing with suppliers is an important issue. Establishing a long-term relationship with a supplier and information sharing with suppliers are commonly used techniques to improve the supply chain performances. Questions of economies to scale and long term commitments are also presented.

The static nature of most inventory management models causes implementation difficulties in a dynamic, uncertain environment. Thus, buffers are used to protect the manufacturing system against shortages of raw materials and supplies resulting from uncertainty in demand and in supply. The concept of buffer management and control is introduced and the effect of such buffers on the performances of the systems explained.

The purchasing function in the Operations Trainer deals with supplier selection as well as inventory management policies. By providing commonly used inventory management models to mange raw material inventories, the Trainer develops insights into the performances of these models in a dynamic environment. Supplier selection is a user's decision in the Trainer. Both lead time and cost are affected by this decision.

The purchasing function in the Trainer is integrated with marketing and production. New opportunities in the market place are presented by the Trainer as requests for proposals to bid on large, sometimes long-range contracts. The availability of raw material, supplier lead-time and availability of production capacity are important factors to consider in analyzing these opportunities.

By integrating purchasing decisions with production and marketing, the Trainer presents the IPOM environment to the user and promotes an integrated dynamic approach to operations.

Chapter 6 - **Scheduling**: Scheduling is an interesting example of a dynamic aspect of operations that is frequently analyzed by static models. The common scheduling technique - implementation of priority rules is usually based on the assumptions that no new jobs enter the production system during the planning period, and resource capacity is a deterministic variable.

The chapter starts with a discussion of priority rules and the performances of such rules in a static job shop. The repetitive production environment where a number of similar products are manufactured repetitively on the same resources is

and can be analyzed immediately. The Trainer provides a framework and a tool for the reengineering of operations processes.

Teaching IPOM to executives and managers helps the introduction of ERP and promotes change through organizational learning. The design of new processes the development of appropriate performance measures and control systems, support the management process in a dynamic, uncertain, competitive environment. Management that understand and practice IPOM has better tools for selecting implementing and using ERP systems.

Problems

1. Explain the tasks of the operations function in a hospital.
2. Describe and explain the five core processes in a fast food chain.
3. Explain the difference between traditional Operations Management and IPOM- Integrated Production and Order Management.
4. Which organizational functions are involved in each process in a fast food chain? Explain the tasks of each function in each process and the interfaces between the different functions.
5. Describe a model used to solve real problems in each of the following areas: marketing, transportation, finance.
6. Select a process in an organization you are familiar with, describe the process and the major decisions required to perform the process. Explain which of the decisions are automated. Explain why it is impossible to automate the ad hoc decisions.
7. Use the concepts of levels and rates and show the relationship between the number of applicants to vacant position, the level of unemployment and other relevant economic factors.
8. Explain the relationship between the flow of material, the flow of information and the decision making process while shopping at the supermarket.
9. Maps are a model of reality show how a real problem can be solved using a map as a model and explain the simplifying assumptions and under what conditions the solution of the problem may not be applicable for the real problem.
10. Find an article about the ERP concept and explain the differences between this concept and traditional Management Information Systems.

2 ORGANIZATIONS AND ORGANIZATIONAL STRUCTURES

2.1. Functional and project organizations, typical goals and performance measures

The history of organizations is probably as long as the history of mankind. Early organizations like families or groups of hunters evolved into tribes, kingdoms and empires. The need to survive in a hostile world, to carry out missions too great for a single person and to share scarce resources are just some of the reasons for the creation of early organizations.

Our modern society and its rapidly developing, complex technology, which results in the specialization of experts in very narrow fields, created an additional reason for the existence of organizations. Most products and services today, are based on the integration of hardware, software, data and human expertise - a combination which a single person usually does not fully master. Thus, organizations in the form of expert teams are created to compete in today's markets.

Organizations designed to produce goods and services are as old as the pyramids in Egypt or the Temple in Jerusalem. Both required the coordinated work of many people in order to be accomplished. The principles of division of labor and specialization are fundamental to many of these organizations. Adam Smith in his book *An Inquiry into the Nature and Causes of the Wealth of Nations* (1776) describes the manufacturing of pins using these principles . The operation that he describes is a result of a well designed process based on division of labor- each person involved repetitively performs a small part of the work required to manufacture a pin thus rapidly becoming very efficient in performing the task assigned to him.

The principles of division of labor and specialization are useful only if a good coordination is maintained between the different components of the organization. Coordination in a highly repetitive environment is relatively easy to achieve, as exactly the same processes are performed repeatedly by each person involved. Thus special approaches such as the synchronized assembly line can be implemented. If the variety of products or services supplied by the same facility is significant, the problem of coordination is much more difficult to handle. The problem of coordination is most difficult when unique products or services are required and very little repetitiveness exists. In this case special attention to the scheduling of resources and activities is required and special techniques for project planning and control are used (Shtub et. al. 1994).

There is a difference between formal and informal organizations. Most formal organizations are based on a clear definition of responsibility, authority and accountability while informal organizations are based on common interests, common beliefs, social values, feelings, tradition, etc., the following discussion focuses on formal organizations.

There are many forms of formal organizations. Few "prototype" organizations are common in business and industry:

The Functional Organization is based on grouping individuals into organizational units according to the function they perform. Thus, individuals dealing with customers and markets are in the marketing division, those responsible for the purchasing of goods and services for the organization are in the purchasing division, while engineers are members of the engineering group.

In large organizations each division is usually subdivided into smaller groups to facilitate better coordination and management. Thus, the engineers may be divided into those dealing with new product development, and those responsible for manufacturing, quality or field service.

The organizational structure of a typical functional organization is depicted in Figure 2-1.

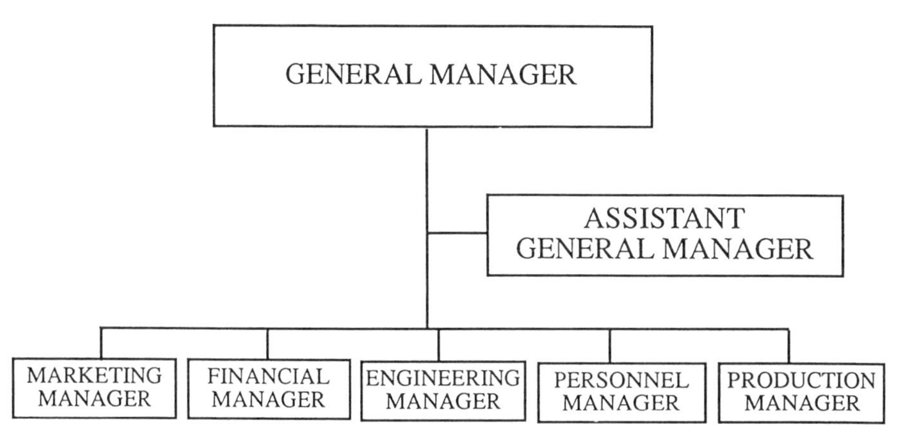

Figure 2.1 The functional organization

An advantage of the Functional Organization is the pooling together of similar resources. By pooling people who share a common expertise and responsibility and providing these groups of experts with suitable facilities and equipment, better utilization of resources is achieved. Furthermore, the flow of information within each part of the organization is made simple due to common background, terminology and interests of the people in each group. A good flow of information is an important factor that affects organizational learning.

Management of the order fulfillment process in a project structure is efficient when each customer order is large enough to justify a dedicated project team with minimum waste of resources. This is the case for example in the heavy industry like shipbuilding.

To overcome the problem of duplicate resources and consequently the low utilization of resources, a combination of the functional structure and the project structure was developed - the matrix structure. In a matrix structure the functional organization is the basis. Project teams are formed when needed by assigning part-time experts from the functional units to the project managers. Thus, most resources are still pooled together while the customer has a single contact point - the project manager. There are many variations of the matrix structure - if the projects are small even the project manager may devote only part of his time to the project while being a member of a functional unit. On the other extreme, in a large, complex project a core project team may be formed with few experts assigned full time to the project. The project team is supported by people from the functional part of the organization on a part time-as-needed basis.

The organizational structure of a typical matrix organization is depicted in Figure 2-3.

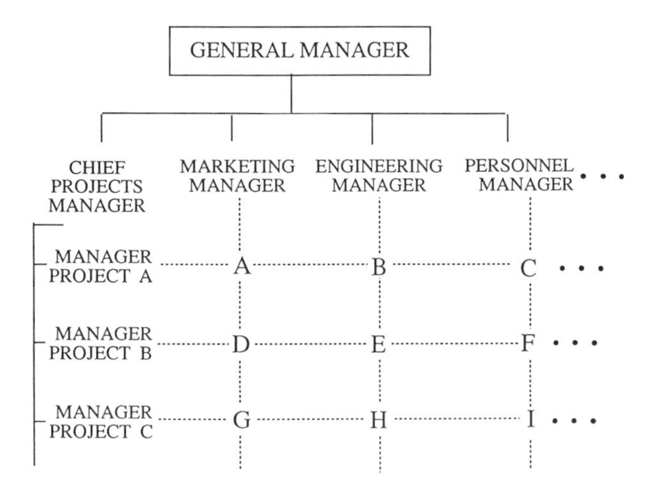

Figure 2.3 The matrix organization

The design of an organizational structure is a difficult yet extremely important task. A structure that was very effective over a long period of time might not perform properly when the environment changes. Thus a order fulfillment process managed successfully in a functional organization serving a stable market may suddenly fail if competition increases ,pushing lead-time down while increasing the pressure for a larger variety of end products.

Consider for example the process of new product development in the above situation. For many years functional organizations used to have special organizational units responsible for new product development. When marketing identified a need for a new product the product development department designed the product and produced all the drawings and related documents that define the physical and functional properties of the product. Manufacturing engineering, based on the product definition developed the processes for manufacturing, assembly and testing of the product and logistics found suppliers for raw materials and components, developed packaging and shipping procedures etc. In recent years this functional oriented process of new product development is severely criticized: it is too slow and it does not provide sufficient "value " to the customers.

The process is slow because of its sequential nature. Each organizational unit is like a link in a chain - it gets its input from the previous unit in the process and produces output, which is the input for the next organizational unit that participates in the process. Since only one organizational unit is involved at a time the duration of the development cycle is the sum of the duration of the processes performed in the participating organizational units.

The problem with a product that provide insufficient value to the customer is generated by important decisions regarding the characteristics of the product that are made in the early stages of the product life cycle (product definition and product development stages). These decisions regarding the product's physical and functional characteristics, have major influence on the product's life cycle cost and therefore on its value to the customer. Since many experts in new product development were not trained to minimize the cost of manufacturing, operating and maintaining products, their decisions represented a local optimum- and frequently resulted in a product design with many functions and options but also very expensive to manufacture, operate and maintain.

In today's competitive markets a fast product development process that yields new products with high value to the customers is a must for survival. In many organizations a new approach to new product development is implemented - Concurrent Engineering (CE). In CE, teams of experts in new product development, manufacturing, operation and maintenance collaborate with the customer to design products with the highest value possible i.e. the performances that the customer wants for the lowest life cycle cost. In CE, experts from the different areas team up in the process of developing the best possible product. CE represent a new organizational structure, which is based on the process, performed by a team .It is different from the project structure and the functional structure as it deals with a specific process: the process of new product development.

An approach similar to CE in new product development is needed for the order fulfillment process. A new organizational structure - a team responsible for the entire order fulfillment process is required. A team of experts in marketing, operations and purchasing that manages the order fulfillment process from receiving a customer order to the delivery of the required goods and services. A team-based approach to the order fulfillment process is the key to success in today's competitive market. The objective of CE is to develop high quality products (quality based competition) with the lowest life cycle cost (cost based competition) in the fastest way (time based competition). A similar objective should be adopted by the team managing the order fulfillment process i.e. to develop and maintain an order fulfillment process that minimizes the lead time from receiving a customer order to supplying it, that minimizes the cost of the process and that yield high quality in terms of eliminating deviations of actual supply dates from the promised dates.

2.2. The job shop, flow shop, and group technology

The organization of people according to the functional areas in the functional organization, according to a mission in the project organization, or according to the process they perform as in CE is in many ways similar to the organization of physical resources on the shop floor. In a manufacturing organization physical resources are machines, tools, inventories of goods, material handling systems, furniture, office equipment, etc. In a service organization furniture and office equipment are also used but machines are of different types (food processing equipment and refrigerators in a restaurant, or x-ray machines and operating room equipment in a hospital). Like people in the organizational structure, physical resources can be organized or laid out in several ways:

The job shop: This is a functional oriented layout where machines and equipment, which perform a similar function, are grouped together. Job shops can process a variety of products by routing each product according to the specific process it requires. Like a functional organization pooling of similar resources is the advantage of the job shop. However, coordination of the manufacturing process of a product requiring processing at different "departments" is difficult, as each product type has a special manufacturing process and consequently a special sequence of operations on the machines.

The layout of a typical job shop is depicted in Figure 2-4.

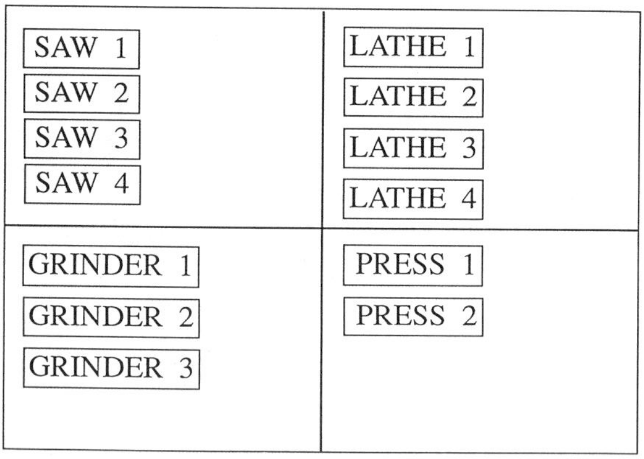

SAW 1 SAW 2 SAW 3 SAW 4	LATHE 1 LATHE 2 LATHE 3 LATHE 4
GRINDER 1 GRINDER 2 GRINDER 3	PRESS 1 PRESS 2

Figure 2.4 The job shop layout

The flow shop: This is a process oriented layout in which products with similar routing or manufacturing processes are manufactured on a dedicated facility in which the layout of machines follows the processing sequence of the products. Product movement in the shop follows the machine layout, which follows the product routing, thus management of the flow shop is made easier.

Flow shops are a natural approach to the facility layout when a family of products with similar processing requirements is manufactured by an organization.

The layout of a typical flow shop is depicted in Figure 2-5.

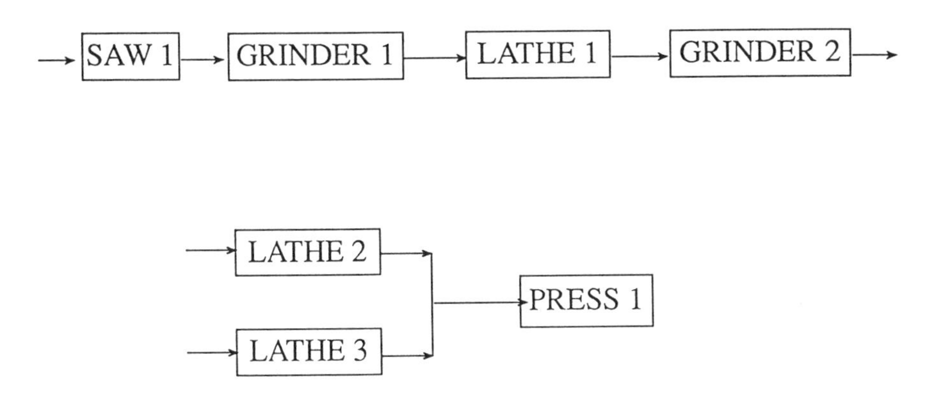

Figure 2.5 The flow shop layout

Group Technology is a technique designed to identify products with similar process requirements and to group these products into families processed by the same equipment arranged in a manufacturing cell. By using the Group Technology concept it is possible to transform a job shop into several flow shops, each processing a family of products that requires a similar sequence of operations. Group Technology is an example of a transformation from a functional structure to a process oriented structure of the physical layout of a facility.

The layout of a typical shop laid out according to the Group Technology principles is depicted in Figure 2-6.

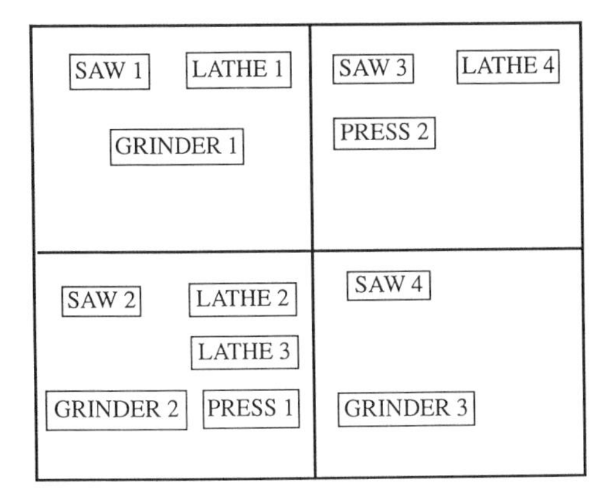

Figure 2.6 The Group Technology layout

When Group Technology principles are correctly applied, the order fulfillment process of the family of product manufactured in a technology cell is managed by the team assigned to the cell. In this case a team responsible for the entire process manages the order fulfillment process. The problem is to implement the same approach when families do not exist. ERP systems provide a solution. It is possible to define "virtual" cells in the model base and to manage these cells based on information provided by the ERP system as explained in Chapter 8.

2.3. Operations management and its interface with other functional areas: restructuring the order fulfillment process

The coordination between functional organizational units performing the order fulfillment process in the functional organization is a difficult task. This is due to the fact that similar goals, terminology and experiences are shared between people in the same organizational unit but not across different units in the organization. To improve the order fulfillment process a solution similar to Concurrent Engineering in new product development and the order fulfillment process in Group Technology is needed. A solution based on assigning the responsibility for the whole order fulfillment process to one team having a common goal to achieve better coordination among everybody involved in the process.

The order fulfillment process begins with the customer. Two pieces of information are needed at this point: The cost of the product, and the delivery lead-time. As discussed in the next chapter, in the functional organization we assume that marketing can receive the first information from the accounting system, while lead-time information is readily available from Operations. In reality, however, both cost and lead-time are changing continuously over time i.e. both are dynamic and their momentary value depends on the current load on the shop floor.

Once a customer order is generated the interface with suppliers is important: purchasing of raw materials and parts, as well as subcontracting (make or buy) decisions are part of the next step in the process. Purchasing people need to know what to order, in what quantities and what is the required due date for these orders. The common assumption, that the Operations function provides a clear-cut answer to these questions may be misleading, as explained later in detail in Chapter Seven. This is due to the basic assumption that lead-time is an input to the operations planning and control information system. In reality this input is a guess and the actual value depends not only on the shop load but also on the very same guess that is used as input. Furthermore, required quantities of purchased parts and material as well as required due dates change continuously over time due to the dynamic nature of operations. Finally make or buy decisions are based on the current and future load on the shop floor- information that is ever changing and usually not known to the purchasing people. Thus, a purchasing department that is not continuously involved in the order fulfillment process may not deliver the right parts and materials in the correct quantity, on time.

The order fulfillment process - from customer needs to manufacturing, through purchasing and subcontracting is subject to uncertainty. Customers change or cancel orders, forecasts of demand are subject to errors, availability of manufacturing resources is subject to machine breakdown and employee absenteeism, while suppliers lead time, quantities and quality are all subject to random variation.

A well-integrated order fulfillment process is required in order to cope with the dynamic, uncertain environment. The organization which supports such a process should not be divided by functional lines and it should focus on one goal and coordinate all of its efforts to achieve this goal - a competitive order fulfillment process that outperform competition in cost, quality flexibility and time.

Integrated Production and Order Management starts with a new organizational structure- the establishment of a management team responsible for the entire order fulfillment process. This team deals with the customers - forecasting demand and managing customer orders, it is also responsible for manufacturing and it is the link to suppliers and subcontractors. A single manager is responsible for the entire order fulfillment process and he performs his job with the help of his team. It is important to locate the members of this team in the same facility and adjacent to each other in order to promote good communication within the team. This concept of an integrated team is difficult to implement in organizations where functional units exist. The Operations Trainer is instrumental in supporting this organizational change by providing a setting where the team approach to IPOM can be experienced and tested.

A well structures order fulfillment process can be implemented once the team is established . This process provides clear answers to the following questions:

- What should be done to perform the process successfully?
- When should it be done?
- How should it be done?
- Who should do it?
- What information should be provided, to whom and in what format?
- What performance measures should be used?

Answering these questions is the process of designing the order fulfillment process.

2.4. Functions and scenarios in the Operations Trainer

The Operations Trainer simulates the entire order fulfillment process. It contains four functional areas integrated into a single system supported by an ERP like information system composed of a model base and a common database. The four functional areas are Marketing, Purchasing, Production and Finance. The first three areas are active in the order fulfillment process while the finance area only provides information and performance measures.

A main screen on which current information on the scenario being simulated is presented and is available in each of the four areas. Thus, in the Production area the main screen provides information on the current status of the work centers including:

The Finance main view is a standard profit and loss (P&L) statement which is updated at the beginning of each month. At the beginning of a new year, last years report is presented.

Each of the four main views is presented within the same framework. Five submenus appear on the menu line of the screen: Session, Information, Action, Policies and Help. These menus are specific to the current main view and will be discussed later. On the left hand side of the screen general information regarding the current session is presented:

- Clock - simulation date and time
- DDP - Due Date performance for the entire scenario - based on the average DDP of the end products.
- Cash - Current cash position of the simulation
- Timer - presents elapsed time since start of simulation
- Status - presents the status of the current run
- Session end - target date for finishing the current run

In the following chapters the role of each functional area in the order fulfillment process will be discussed as well as the implementation of this role in the IPOM environment including the detailed description of the relevant screens in the Operations Trainer.

The Operations Trainer supports the traditional functional organizational structure, and it can be used by a group of people each acting as a functional area. It is designed to simulate the IPOM environment as well, this is done by asking the users to work as a team making all the decisions regarding the order fulfillment process together.

The Operations Trainer can simulate any given scenario, to support the discussion on flow shops and job shops in this chapter files containing these two scenarios are available with the trainer.

Problems

1. Select an organization that you are familiar with and draw its organizational chart. Explain the tasks performed by each function, its goals and performance measures.
2. Discuss the application of division of labor in a restaurant and in a supermarket.
3. Explain the difference between the formal organization and the informal organization in a university.
4. Under what conditions is the project organizational structure most appropriate? Give an example.
5. Write a job description for a project manager for a Research and Development project and a job description for a manager of an engineering department in a matrix organization. Explain the goals and responsibilities of each position.
6. Find an article about a successful organization. Based on the article explain how success is measured and assessed.
7. Consider the layout of a typical hospital. Is it similar to a job shop, a flow shop or a group technology based layout?
8. Explain the interfaces between the operations manager of a fast food chain and other functional managers. Focus on the flow of information and on the decision making processes.
9. Discuss the pros and cons of group decision making. Under what conditions is group decision making the preferred approach? Give an example where group decision making is preferred and a counter example where group decision making is not recommended.
10. Use democracy as an example of group decision making. Discuss the different levels of group decision-making based on this example.

3 INFORMATION AND ITS USE

3.1. From data collection to decision making

The order fulfillment process as defined earlier - from the acceptance of a customer order to the delivery of the right product, in the right quantity, at the right time to the customer has three integrated aspects:

- the data aspect - data collection, storage, retrieval and analysis
- the decision-making aspect - making decisions regarding the usage of information, resources and materials.
- the physical aspect - handling and processing of material (raw material, parts and finished products)

The first aspect deals with the generation and collection of data such as a new customer order that enters the system. New data is generated when transactions are taking place, new data is also collected from a variety of internal and external sources by data collection systems.

Data can be classified based on its importance and urgency. Urgent data may have to be processed, analyzed and made available to the decision makers immediately, while important data may wait in storage from where it can be retrieved for further processing and analysis when needed. Data classified as unimportant should not be collected in the first place and if collected it should be discarded. The technical aspects of data storage and retrieval are discussed in detail in books dealing with IT-Information Technology and IS-Information Systems (see for example Kenneth and Jane Laudon 1995).

Data is used to support decision-making. There are two types of decisions associated with the order fulfillment process:

Routine decisions these decisions are based on the same type of data processed routinely in the same way. An example is decisions regarding the payment to suppliers. Given the cost of goods supplied and agreed upon, the quantity supplied, and the payment agreement (say thirty days after delivery) the decision of how much to pay and when, is routine and can be automated.

Ad - hoc or non-routine decisions these decisions are difficult or impossible to automate. A decision to discontinue a product line, to change the quoted price of a

product or to give a special, one time, discount to a customer, are examples of non routine or ad - hoc decisions.

It is not always simple to distinguish between the two types of decisions. One guideline is the amount of intuition, experience and emotions involved in the decision making process. Another point to consider is the frequency in which each decision is made. Frequent decisions that are based on well defined (and preferably quantifiable logic) are candidates for automation, while decisions that are very rarely made and are largely based on the decision maker's intuition and experience are less suitable for automation.

One attribute of a poor order fulfillment process is the lack of automatic decision making. When this is the case each decision is treated as one of a kind and top management is spending most of its time dealing with an endless stream of routine decisions that should have been automated in the first place. In a well-designed order fulfillment process, routine decisions are made by the information system in an automatic mode. In case a routine decision can not be automated for some reason, an effort should be made to transfer such a decision to the employees who execute the decisions. They should be instructed on how to make the routine decisions, and should be supported by providing all the information and applicable decision rules (the appropriate analysis of information required to make the decision).

In a well-designed order fulfillment process only non routine, special decisions are handled by top management. Top management should be busy monitoring the environment and the performances of the process in an effort to identify the need for changes in the current process early on. When decision making is automated, it is important to watch the resulting performances by means of a monitoring and control system designed to alert management when a currently used decision making process produces poor results. This is especially true when competition takes place in a rapidly changing environment with high levels of uncertainty.

An important step in the design of the order fulfillment process is to decide which of the decisions to automate, how to do it and how to monitor automated decision making processes to detect poor performances, and the need to make ad hoc decisions or to modify an existing process. The design of the decision-making aspect in the process should therefore focus on the following:

- the ability of the information system to support automatic decision making
- type of decisions to automate
- the logic used for automated decisions
- data required for automated and non automated decisions
- the way the data should be collected, processed and presented as information to support decision making processes
- the establishment of monitoring and control systems that are needed to detect problems in the process early on

When the needed data is collected, processed and supports decision making, the third aspect of the order fulfillment process -the physical aspect- processing and handling of material is reduced to understanding and implementing the decisions, which is mostly a matter of information sharing and proper training for the workforce.

Information sharing is a key element in redesigning the order fulfillment process. Unlike the typical Functional organization in which each function has its "own" data, in the Integrated Production Order Management approach, with the support of the ERP systems, data is shared by everybody involved in the process. Thus, when decisions are implemented those responsible for the implementation understand not only what they have to do but also why it has to be done.

3.2. Information systems - the data base and the model base

Three types of information systems support the order fulfillment process:

- Transaction processing system - a system that performs and records all routine transactions such as sales order entry, inventory transactions and shipments to customers.
- Management Information system - a system that serves the functions of planning, decision making and controlling, by providing the output of automated decision processes as well as routine summaries and exception reports.
- Decision support system - a system that combines data and analytical models to support semistructured and unstructured decision making.

All three systems use data that comes from internal and external sources. An important component of a well-designed information system is a database in which the data is managed and stored. The data base is managed by a Data Base Management System (DBMS) whose most important feature is its ability to separate the logical and physical views of data so that users can retrieve data without being concerned with its physical location in the data base. There are three important types of logical data base systems: Hierarchical systems that are high in processing speed but limited in flexibility, Network systems and Relational systems that are very flexible in supporting ad - hoc requests for information but are relatively slow. The design of data bases is beyond the scope of this book. This important topic is discussed in books on Databases and Information systems.

The model base that supports the order fulfillment process includes three types of models:

- Models for well-structured, routine, decision-making processes - these models can be completely automatic. For example a decision to reorder raw material when its inventory level reaches the calculated reorder point. In some applications the information system may automatically issue a purchase order while in other applications the system only recommends and the order is issued manually by low level management. These models are typically part of the Management Information System.

- Models for non-structured or non-routine problems - in this case the output of the model does not provide a decision that can be implemented but rather the result of the analysis provides some insight into a situation. Models for capacity planing, for example, are used in the order fulfillment process routinely to estimate current and future load on resources. These estimates are helpful for identifying overloaded resources (bottlenecks), modify delivery schedules, outsourcing (using subcontractors) etc. These models are typically a part of the Decision Support System.
- Models for process control - these models are designed to alert management in cases where the process gets out of control. An alert that some customer orders might not be ready for shipment on time, an alert that a work center is over loaded or an alert regarding expected shortage of raw material are typical examples.

Models are used when a real problem is too complicated to analyze and to solve. To facilitate analysis and solution, simplifying assumptions are made and a model is developed. Thus, a good model is simple enough to facilitate understanding and analysis of a real problem and yet it is close enough to the real problem so that the solution obtained from the analysis of the model is useful for the real problem. To test the applicability of a solution obtained from a model, sensitivity analysis is conducted where the effect of the simplifying assumptions is tested.

In the past functional units in the organization tend to develop dedicated information systems. Thus, the accounting information system supported monetary transactions and cost accounting analysis while the production planning and control information system focused on the management of materials and resources. The need for integration and the understanding that local decisions made by functional units affect other functional units promoted the development of integrated information systems. These systems, known as Enterprise Resource Planning (ERP) Systems, integrate the data of the whole organization in the single data base, and provide models for decision support that consider the processes performed by the whole organization to achieve its global objectives. These systems are designed to support all the processes in the organization:

- The development process - from an idea for a new product or service to a working prototype.
- Preparation of facilities - from a working prototype of a new product or service to the successful completion of design, implementation and testing of the production/assembly or service facility and its supporting systems and resources.
- Sales - from the study of the market and its needs to the reception of a firm customer order.
- Order fulfillment - from a firm customer order to the delivery of the required products or services and payment by customer.
- Service - from customer call for a service to the fixing of the problem and a satisfied customer.

An important consideration in the design of an information system that supports the order fulfillment process, as well as the other processes the organization performs, is which models to use in the system. The models are separated from the database so that the same model can be used to solve different problems by applying it to the relevant data in the database. The collection of models in the information system comprises the model base of the system. Another important aspect of the design of an information system that supports the order fulfillment process is the integration between the model base and the database. Ensuring that all the required models are in the model base and all the data required for the selected models is in the database and can be retrieved easily by the appropriate models is of utmost importance.

When an organization develops its own information system to support its processes it can select the models and DBMS most appropriate for its needs. Due to the high cost of developing and maintaining a tailor made, one of a kind, information system, most organizations tend to purchase off-the-shelf software and set it up to fit their needs as much as possible. However, commercial software which is typically less expensive to purchase and maintain, has limited flexibility. Therefore, it is extremely important to design the processes prior to purchasing the supporting software and then to select the software product that fits the proposed processes best. If this is not done in the above sequence and the software is purchased before the processes are designed, limitations of the software may constrain the design of the order fulfillment process and a real dynamic integrated process may not be possible. This is especially important in selecting ERP type system that supports all the processes in the organization. The danger is that system selection is based on the need of one process while the other processes are not well supported by the ERP system.

There are many different software packages on the market that support the order fulfillment process. Some are designed to support parts of the process such as packages using finite capacity scheduling logic for shop floor control, while other packages are supposed to support the whole order fulfillment process, such as advanced Material Requirement Planning packages known as MRP II. Enterprise Resource Planing -ERP systems are designed to support all the processes performed by the organization simultaneously. In the Operations Trainer the whole order fulfillment process is simulated and managed by the user who is supported by an information system. Thus, a user not familiar with the concept of MIS, DSS, MRP, ERP and the use of models to support decision-making can exercise these concepts by using the Trainer. Furthermore, the Trainer provides a laboratory like setting in which the team responsible for the order fulfillment process can design its process, test it, and fine-tune it to fit the needs, the environment and the culture of the organization.

3.3. The accounting information system

The most common information system is probably the accounting system, which is a specialized MIS, designed to collect, process, and report information related to financial transactions. Legal, tax and reporting requirements, as well as the need to manage and control scarce financial resources of the organization, promoted the use of accounting systems in organizations. Over the years these systems were extended to support decision making and performance measurement. Cost accounting is a collection of models commonly used to support decisions on pricing, outsourcing and purchasing.

Following Gleans et al. (1993) the major components of the accounting information system are:

1. The order entry, sales entry system - the system that interface with the customers and the markets.
2. The billing, accounts receivable, cash receipts system - the system that bills customers, monitors customer accounts, and records cash receipts.
3. The purchasing account payable, cash disbursements system - the system that supports purchasing of goods and services, monitors open payable, and processes and records of cash disbursements.
4. The inventory system - the system that monitors inventory and its value.
5. The human resources management system - the system that maintains employee and payroll records and prepare and records payroll transactions.
6. The general ledger, financial reporting system - the system that maintains the general ledger and prepares accounting reports.
7. The integrated production system - the system that collects, processes and records production costs.

The accounting information system is a very important source of information for the management of the order fulfillment process. However, the information generated by this system should be well understood to avoid misinterpretation and erroneous decisions. For example traditional cost accounting models tend to assume that mainly production volume, direct material and direct labor influence costs. This assumption was valid when the level of automation on the shop floor was low and direct material and direct labor costs were the major cost components of a manufacturing organization. Furthermore, some cost accounting models are based on the assumption that direct labor is flexible, and its level of usage can be easily changed with the production volume. Today with the advance manufacturing technology direct labor costs are frequently just a fraction of the total cost of a product or a service and it is pretty much fixed, at least in the short range. Consequently, the analysis of traditional cost accounting models may be misleading. Newer approaches to cost accounting are based on a different set of assumptions. One example is Activity Based Costing (ABC) which is based on a minute analysis of the activities performed in a process. Both value added and none value added activities are accounted for and the system assigns the costs of activities to products based on the actual time each activity is performed on each product and the actual cost of resources used to perform the activity.

A second example is the system of three financial performance measures suggested by Goldratt and Fox (1986):

- Throughput (T)- The rate at which the system generates money through sales. It is the difference between the dollar sales for the period and the expenses generated by these sales (the expenses that would not occur if the sales were canceled).
- Inventory (I) - The entire amount of the money the system invests in purchasing items the system intends to sell. This is the original purchase cost of raw materials, parts and components stored in the system.
- Operating Expenses (OE)- The entire amount of money the system spends in turning inventory into throughput, which is not directly dependent on the level of sales (i.e. not included in the throughput calculations).

The operations Trainer is equipped with a traditional cost accounting system that performs the functions mentioned above. Unlike the information systems supporting the other functional areas in the Trainer, the accounting information system does not support actions and policies developed by the user. Its only function is to collect data, analyze it and provide the user with financial and cost accounting information. Understanding and using this system to support the management of the order fulfillment process is an important part of teaching the IPOM concept.

3.4. Quality of information

Earlier in this chapter we categorized data as urgent, important and unimportant. The information generated by processing the data has a value only if it is useful in the decision making process. The usefulness of information is the result of several qualities and it can be measured by the following:

- Understandability: Information can support the decision making process only if the decision-makers understand it. Thus, if the information is in a language not understood by the decision-makers its usefulness is doubtful. As an example consider an organization in Russia that is interested in implementing a new ERP system. A major factor in selecting the new system is whether a Russian version is available. If a candidate ERP system supports only the English language, the information it provides might be useless unless every user understands English.
- Validity: Valid information describes an actual and relevant reality. Thus, if a production lot is specified in a work order, and the whole lot is rejected after processing due to poor quality, the information that the lot is completed is not valid as none of the units is of accepted quality.
- Relevance: Information is relevant for the decision making process if it reduces uncertainty and either reinforces the decision that would have been made without that information, or change the outcome of the decision making process. For example information about the development of a new version of the ERP system that is used by the organization is irrelevant to the scheduling decisions currently made on the shop floor. The very same

information is very relevant to a decision on whether to replace the current ERP system or not.

- Accuracy: This is a measure of the level of agreement between the information and actual reality. For example, if according to the inventory records there are fifty units of a finished product in stock, but actual counting reveals that there are only forty-five units on the shelf, the information in the inventory records is not accurate.
- Completeness: Measures the level of coverage of every relevant object by the information. For example, if there are one thousand and one different items in inventory but the information system contains records for only a thousand of them, then even though each record might be perfectly accurate, the information in the inventory system is not complete.

A major aspect in the design of the information system that supports the order fulfillment process is to ensure that the information is understood, valid, relevant, accurate and complete. The way to do this is to perform a thorough analysis during the design stage of the process, to understand the needs for information and to develop a cost-effective system to support these needs. The information system should be managed throughout its life cycle to ensure that its performances do not deteriorate over time.

The information provided by the Operations Trainer is accurate and complete, however, part of the experience with the Trainer is to understand the information provided and to assess the validity and relevance of that information.

3.5. Forecasting

An ideal information system provides only high quality information that is understandable, valid, relevant, accurate and complete. In reality some information is simply not available and the best we can do is to estimate it. This is particularly true when future events are considered. For example information about future interest rates, future inflation rates and future money exchange rates is important for investment decisions. Information about future demands and future competition is important for marketing decisions and information about future absenteeism of employees and future breakdown of machines is important for production planning. However, exact information of this type is simply not available and the best we can do is to try and forecast it.

Forecasting is an art as well as a science. Some people can forecast future events with a remarkable accuracy, based on their experience and intuition. For example, some farmers are known to forecast the weather with pretty good results without using any sophisticated models, computation or computers. In this book, however, we concentrate on forecasting models that are frequently used to support the order fulfillment process, and we will focus on the techniques that are embedded in the Operations Trainer. These techniques are based on time series analysis, i.e. an attempt to forecast the future based on the analysis of past data. The future value of a time series is called a forecast while past values are called observations. A time series is depicted in Figure 3-1.

random from a homogenous population), providing no trend, seasonal or cyclical components are present. The most recent observations over the same number of periods are averaged whenever a new forecast is required. Since every time a new observation is available the oldest observation is taken out of the averaging process, this model is known as the "Moving Average" forecasting model.

A formal presentation of the moving average model follows:

$$F_t = \sum_{i=1}^{n} O_{t-i} / n$$

In this formulation F_t is the forecast for period t

O_{t-i} is the observation in period t-i

n is the number of periods in the average

Example

Consider for example the following time series

Jan	Feb	Mar	Apr	May	Jun	Jul	Aug	Sep	Oct	Nov	Dec
32	31	29	30	31	28	29	33	29	27	31	32

Assuming a moving average over 5 months, the forecast for next January is:

$$F_{13} = \frac{32+31+27+29+33}{5} = 30.4$$

A variation of the moving average model is used when a trend is detected in the data. In this case recent observations are more effected by the trend and therefore are considered more important in the averaging process. Thus, if n observations are used, instead of assigning each one an equal weight of 1\n as in the simple moving average model, a higher weight is given to recent observations and a lower weight is given to older observations, keeping the sum of these weights equal to 1. This model is known as the weighted moving average model.

Assuming the following weights for the previous example *0.1, 0.15, 0.2, 0.25, 0.3*

The forecast for the January is

$$F_{13} = 0.1 * 33 + 0.15 * 29 + 0.2 * 27 + 0.25 * 31 + 0.3 * 32 = 30.4$$

The moving average model follows:

$$F_t = \sum_{i=1}^{n} w_{t-i} * O_{t-i} \qquad \text{Where } \sum_{i=1}^{n} w_{t-i} = 1$$

In this model F_t is the forecast for period t

O_{t-i} is the observation made in period t-i

W_{t-i} is the weight assigned to the observation of period t-i

A difficulty with the weighted moving average model is to find the set of weights that yields the best forecast. There are infinite combinations of different weights. A weighting scheme that simplifies the process of weights generation and selection is needed as well as a performance measure to evaluate the quality of different forecasts.

The exponential smoothing model is a weighting scheme that simultaneously generates all the weights for a time series so that recent observations are given heavier weights while the total of all the weights is always equal to one. The fundamental relationship of the exponential smoothing model is:

$$F_t = F_{t-1} + \alpha(O_{t-1} - F_{t-1})$$

Where F_t is the forecast for period t

F_{t-1} is the forecast made for the prior period

O_{t-1} is the observation in the prior period and

α is a smoothing constant.

By substituting F_{t-1} into the equation for F_t it can be shown that the sum of all the weights equal to 1.

Consider the previous example, using a smoothing constant of .5 the following forecast is obtained based on the last five observations:

$$F_t = 0.5 * 32 + (0.5)^2 * 31 + (0.5)^3 * 27 + (0.5)^4 * 29 + (0.5)^5 * 33 = 29.96$$

To select the best value of α a performance measure is defined and a search for the constant α that yields best value of the performance measure is conducted. The smoothing constant that minimizes the performance measure is selected for the exponential smoothing model of that time series.

One performance measure used to evaluate the quality of forecasts is the Mean Absolute Deviation or MAD. The MAD value is calculated as follows:

$$MAD = \sum_{i=1}^{n} \frac{|forecast_i - observation_i|}{n}$$

An alternative performance measure is the Mean Squared Error (MSE). The MSE value is calculated as follows:

$$MSE = \frac{\sum_{i=1}^{n} (forecast_i - observation_i)^2}{n-1}$$

Advanced forecasting packages are designed to search for the best smoothing constant given a performance measure such as the MAD or MSE. The selected smoothing constant is used to calculate the weights of all past observations in the forecasting model.

In the Operations Trainer demand forecasts as well as capacity requirements forecasts are based on the exponential smoothing model. These forecasts are available to the user as part of the decision support system embedded in the Trainer.

3.6. The accounting information system and financial aspects in the Operations Trainer

The Accounting Information System in the Operations Trainer features all three components of information systems mentioned earlier in this chapter:

- Transaction processing system - this automatic system performs and records all routine transactions including sales order entry for small customers and contracts for clients, inventory transactions for raw material and finished goods inventories, shipments to customers and clients. Each scenario specifies the billing arrangements for suppliers, clients and customers. The system automatically issues and collects payments according to the specific scenario. Cash transactions are also automatic and the cash position is continuously displayed to the users.
- Management Information system - this system provides on line information on the financial situation of the simulated organization. A Profit and Loss (P&L) statement is updated every month. A cost accounting system that serves the functions of planning, decision making and controlling provides reports on product cost analysis and contract cost analysis. This information combined with a list of active contracts supports routine marketing decisions.
- Decision support system - this system provides some special analysis to support ad-hoc decisions. When a new contract is proposed a cost analysis of this contract is performed to support the decision on whether to accept or to reject the proposal. The three performance measures suggested by Goldratt and Fox (1986) in the book "The Race" are also available.

The Economic Value Added (EVA) for the simulated organization is also calculated. EVA is defined as the difference between the net profit made and the interest that could have been generated if all the capital invested in the organization was invested instead in a bank account for the current interest rate.

The accounting information system of the Trainer represents a typical system of this type and provides the basis for a discussion on the pros and cons of such systems.

Problems

1. Describe the order fulfillment process in a Pizza restaurant that has a dining facility and also makes home deliveries. Explain the differences between the two types of service focusing on the aspects of decision-making, flow of information and flow of material in each service.
2. In problem 1 discuss the possibilities of automating the decision-making processes of each service.
3. What kind of data is handled by the transaction processing system of a hospital?
4. Describe a Decision Support System that you are familiar with and explain one of the models used in this system.
5. Why is it important to separate the database from the model base in the ERP environment?
6. Select an organization and describe its accounting system's seven major components, based on Gleans et. al. approach. Explain each component.
7. Under what conditions will Throughput and Sales be exactly the same?
8. Study a map of the city where you live and discuss the quality of information presented in the map.
9. Based on the inflation rate in the last ten years forecast the rate of inflation for next year using three different forecasting techniques. Using MAD as a performance measure, select the best forecasting method and explain your analysis.
10. Study scenario TS100 in the Operations Trainer and discuss the situation of the company at the start of the simulation based on its Profit and Loss statement.

4 MARKETING CONSIDERATIONS

4.1. Make to stock, make to order, assemble to order policies

Integrated Production and Order Management (IPOM) is an integrated order fulfillment process that crosses the traditional functional and organizational lines. This process that starts with customer orders for end products (in the following discussion the term end products will be used for products or parts supplied to customers), proceeds with purchasing from suppliers and subcontractors and includes production, assembly, packaging and shipping of the finished products. The whole process is aimed at fulfilling customer's orders to ensure the long-term success of the organization. Since the process is triggered by customer orders, and is designed to fulfill such orders it is important to define exactly what constitutes an order that triggers the process.

A customer order may be a formal document i.e. a written form specifying the goods ordered, quantities, the required due date, terms of payment and the customer legal obligation to pay, it might be a long term contract with periodical (daily, weekly) deliveries, or, it might be an order issued by marketing based on a demand forecast. In the ERP environment these are all electronic transactions that are aggregated in the Master Production Schedule into a master plan as explained later in this chapter. The selection of the triggering mechanism that drives the whole order fulfillment process is an important decision as it defines the interface between the market and the rest of the process. This decision affects the whole order fulfillment process.

Competition in the order fulfillment process has several dimensions: time, cost, flexibility and quality. The performances of the organization in each of these dimensions have a significant effect on its competitiveness and on its ability to survive. A competitive market that demands low cost, high quality products in a short lead time, coupled with a flexibility to respond to changing needs of the customers, forces management to continuously seek ways to improve the order fulfillment process, to ensure the long term success of the organization.

The design of the order fulfillment process should focus on each of the above dimensions and on the interactions between them. Thus the way production and purchasing orders are initiated is a key design factor in the order fulfillment process. As discussed earlier, the process can be triggered by a firm customer order or by a forecast of future demand. When the trigger is a forecast of future demand, inventories of finished goods are frequently used to buffer against the uncertainty caused by forecasting errors. The promised lead time to the customer is relatively short as it is determined by the time required for some information processing (order processing) and the shipping time from the storage to the customer. This promised lead time is shorter than the lead-time that is required in the case that the process is triggered by firm customer's orders. In the later case the lead time promised to the customer should be long enough to include not only order processing time and shipping time but also the actual manufacturing time of the goods ordered and sometimes the time required to get raw materials from suppliers as well. The competitive advantage of the first alternative - the ability to supply in a shorter lead time, does not come for free- as we saw in the previous chapter forecasts are subject to forecasting errors. Forecasts based manufacturing orders generate excess inventories whenever the forecast exceeds actual demand, if, however the forecasts turn out to be lower than actual demand, the results are shortages, and consequently late deliveries. This interaction between the dimensions of time and cost is further complicated since the buildup of inventories reduces the flexibility of the organization to modify the design of its products, to introduce new models to the market and to change the production schedule when a customer requests it.

In most organizations manufactured goods are made of purchased parts and raw materials which makes the triggering decision even more complicated. Management can decide to trigger the purchasing activities based on a demand forecast, but to start production and assembly only when firm customer orders exists. The designer of the order fulfillment process has several alternatives to initiate purchasing orders and production orders, for example:

- issue purchasing orders and production orders based only on firm customer orders.
- issue purchasing orders and production orders based on demand forecasts, maintain inventories of some purchased raw materials and parts.
- issue purchasing orders based on the current inventory of these purchased parts and raw materials and issue production orders based on firm customer orders.

The number of different alternatives is large as each part or raw material may be managed differently according to its cost, lead-time, and the end products for which it is being used. Again the trade-off between the dimensions of lead-time, flexibility and the cost of inventory emerges.

The above discussion brings up the need for a well-integrated order fulfillment process: conflicting marketing, production and purchasing considerations create the dilemma of how to trigger production and purchasing orders. Without a well-defined process for triggering new orders, every order is an ad -hoc decision that requires management attention.

There are extreme policies for triggering new production and purchasing orders. One such policy is stocking enough finished goods to guarantee off-the shelf delivery of every existing end product, the opposite approach is "no stock production" (Shigeo Shingo, 1983) according to which manufacturing and

To avoid this conflict the whole order fulfillment process should be integrated to achieve a common set of goals with respect to cost, time, flexibility and quality. Appropriate performance measures should be developed and applied to the order fulfillment process to support its management and control. The MPS is an integrating device in the order fulfillment process. By making the MPS known to all the members of the IPOM team and by introducing a team approach to the management of the MPS a common goal for the order fulfillment process is established, maintained and made known to the whole team.

The management of the MPS consists of three activities:

- the introduction of new requirements into the MPS,
- the updating of existing requirements - changing the required time or the required quantities, or changing the status from available to promise to a firm customer order,
- monitoring and control of actual performances of the order fulfillment process compared to the goals established by the MPS.

Each of these activities require an integrated approach: The introduction of new requirements into the MPS should be based on the market demand, on the current and future load on the shop floor and on the availability of purchased parts and raw materials. Changes in existing MPS records should serve the market needs subject to capacity and material availability constraints. The monitoring and control of the MPS should produce early warnings regarding late manufacturing or purchasing orders that may result in a delay in supplying the planned gross requirements. Such early warnings can support the order fulfillment process management team in setting priorities, expediting orders or utilizing additional sources of capacity such as overtime, additional shifts and subcontracting.

The MPS represents the gross requirements for end products, Figure 4-1 depicts a schematic MPS:

PEROID / GROSS REQUIREMENTS OF PRODUCT:	1	2	3	4	5	6	7	8
A	50	–	–	32	30	65	17	77
B	30	45	–	94	–	93	44	24
C	–	30	66	62	20	–	–	86
D	25	75	28	53	–	31	56	–

Figure 4.1 A schematic Master Production Schedule (MPS)

Translation of the MPS into work orders and purchasing orders that in turn trigger the material handling and processing activities of the order fulfillment process are performed by Management Information Systems such as MRP or production planning and control systems such as JIT as explained in chapters 6 and 7.

4.3. Lead time and time based competition

Time plays an important role in the order fulfillment process. As explained in the previous section, a shorter lead-time promised to customers improves the competitiveness of the organization and its ability to get new business. A reliable order fulfillment process that consistently deliver on time maintains customer satisfaction and loyalty. The relationship between the lead time promised to customers and the actual lead time of the order fulfillment process is a key factor in the decisions on how much inventory and what kind of inventories (raw materials, parts, or finished goods) should be carried. Thus a continuous effort to reduce the lead-time of the order fulfillment process is critical for developing and maintaining the competitive edge of an organization.

The lead time of the order fulfillment process is determined by the time required to perform the three aspects of the process namely data processing, decision making and the physical processing and handling of materials, parts and finished products.

The order fulfillment process is a collection of operations dealing with data processing decision making and material processing and handling. The duration of each operation, delays before, after and during operations and the ability to perform operations in parallel are key factors affecting the total lead time of the order fulfillment process. Thus to reduce lead time, each operation related to data processing, decision making and physical handling and processing of material should be studied and optimized. Furthermore the process as a whole - the sequence in which the individual operations are performed should also be analyzed to minimize the total duration of the process. This effort should concentrate on four points:

1. elimination of unnecessary operations
2. minimization of the duration of necessary operations
3. minimization of delays before, after, and during the operations
4. minimization of dependencies between operations to enable parallel operations

The first step calls for a critical review of each operation in the process asking WHY is the operation performed and is it necessary. The crucial question is what is the value added of each operation. Operations that do not add value to the customer should be eliminated. Typically Move operations and Storage do not add any value to the customer and should be eliminated or at least reduced to a minimum by selecting a proper layout that follows the processing sequence of the products and minimizes the need for material handling. Similarly manual copying of data double entry of data and mailing of hard copies should be eliminated. Operations that add value to the customer should be combined if possible to eliminate transportation and delays between such operations. For example quality control operations adds unnecessary time to the order fulfillment process. An effort to develop stable, reliable processes and to integrate quality control activities within each operation so that each operator is responsible for the quality of his work can substantially reduce the time of a process.

The next step concentrates on the question HOW value added operations are performed? is it possible to reduce the time and effort or to automate these operations?. Time and motion studies are basic means in this analysis. An effort to develop tools, jigs and fixtures that reduces the duration of operations related to material handling and processing is part of this step. A simultaneous effort to improve data processing and decision making activities by developing appropriate management information systems and decision support systems can reduce the lead time of the order fulfillment process further.

In the third step of the analysis the focus is on delays. Delays are found in all aspects of the order fulfillment process, two types of delays are common: operations delays and process delays.

Operations related delays are generated when due to relatively long set up time large batches (or lots) of the same product are processed. When operations are performed on one unit of the product at a time the whole batch is delayed until every unit is completed i.e. the delay is equal to the set up time plus the processing time per unit multiplied by the batch size. To eliminate this delay an effort to reduce setup time is required. A short setup time makes large lots unnecessary and the delay is eliminated. If setup time reduction is not possible the batch related delay can still be reduced by transferring single product units from operation to operation not waiting for the completion of the whole batch. Thus, by reducing the set up time and the size of the processing batches along with an effort to use one unit transfer batches, the operation delays can be minimized.

Process delays are caused by bottlenecks in the process. A bottleneck is created when the load on a machine is close to or exceeds its available capacity. In this case if operations feeding the bottleneck generate output at a rate higher than the bottleneck production rate a queue of batches awaiting processing is accumulated in front of the bottleneck and cause a process delay. To eliminate this type of delay the bottleneck production schedule should serve as a driver for the schedule of all other operations. The scheduling technique called Drum, Buffer, Rope (DBR) is based on these principles and discussed in Chapter 6. Another possibility is to use finite capacity scheduling techniques that schedule production according to the available capacity of the resources on the shop floor.

In data processing and decision making processes delays are introduced when data and information are made available to users sequentially. In the days that information and data were transferred by means of written documents, the transmission of data was performed by physical transfer of these documents. The time to circulate documents introduced a necessary delay in the process. Today's ERP technology supports the parallel distribution of data by computer communication to eliminate this type of delays.

Decision making delays - delays in the decision making process are common to functional organizations in which information is transferred through the different hierarchy levels in the form of documents and the formality of the decision making process require management signatures on the documents.

Decision making delays are generated in organizations that do not have well-established policies regarding repetitive decisions. In this case many decisions are treated as one time, ad-hoc decisions that require special management attention. The sheer number of such decisions and the need to reach consensus among different organizational units involved in the order fulfillment process translates frequently into an endless number of meetings in which group decision making is taking place. Scheduling such meetings and reaching consensus is time consuming and decisions are often delayed and delay the whole process.

The introduction of control points in the data processing and data distribution process is also a common source of operations delays. When control is exercised by regulating the flow of information it delays the flow of information and forces a sequential decision making process.

Data acquisition causes further delays when it requires manual data entry. Sometimes the entry of the same data several times in the process is the source of delay (and errors), by introducing electronic data requisition and computers communication technology, data acquired anywhere in the process can be made available to all users simultaneously.

The last step in the analysis focuses on the question WHEN each operation is performed in the process. This step concentrates on the longest sequence of operations in the process (analog to the critical path analysis in a project - see Shtub et al. 1994) and on the duration of setups. The objective of the analysis is to reduce the duration of the longest sequence by checking the precedence relations among its activities and wherever possible to schedule activities in parallel. Reduction of the duration of setups is of utmost importance as shorter setups enables smaller batch sizes thus minimizing delays and reduce the duration of the whole process.

By adopting a dynamic team approach to the order fulfillment process including data sharing, elimination of non value added operations and delays, parallel operations, and a consistent effort to reduce set up and operations times the duration of the order fulfillment process can be compressed to provide a competitive edge in the time based competition.

4.4. Quality and its management - quality based competition

The order fulfillment process is the link between the organization and its customers. The successful translation of customer's orders into the actual delivery of goods and services is measured by customer's satisfaction. Thus quality management in the order fulfillment process is the continuos effort to please customers and to achieve customer satisfaction.

Customers are satisfied when their expectations are met or exceeded. These expectations are product and service related. David Garvin (1987) suggested that quality based competition has eight dimensions or performance measures:

Consider a production line used for three models of the same product A,B and C. The direct labor and material cost data are summarized in Table 4-1

Table 4-1. Direct cost data

	A	B	C
Annual demand [units]	100	200	400
Direct labor [hours/unit]	10	5	5
Hourly rate [$]	20	20	20
Direct labor cost/unit [$]	200	100	100
Direct material cost/unit [$]	100	100	100

The annual indirect cost of the line is $80000, allocating this cost among the three models based on direct labor hours and assuming a selling price of $400 per unit of each model the following unit cost and profit are calculated and summarized in Table 4-2

Table 4-2. Total cost and profit

	A	B	C
Total direct hours	1000	1000	2000
Indirect cost/direct hour [$]*	20	20	20
Indirect cost/unit [$]	200	100	100
Total unit cost [$]	500	300	300
Unit selling price [$]	400	400	400
Profit per unit [$]	(100)	100	100

$$* \text{ Indirect cost/direct hour [\$]} = \frac{80000}{1000+1000+2000} = 20$$

Based on the above analysis total profit from the three models is

-10000+20000+40000=$50000

Since model A is losing money it seems that the right decision is to stop its production.
In this case the annual direct cost of production is calculated as follows:

Labor cost	100*200+100*400=$60000
Material cost	100*200+100*400=$60000
Indirect cost	$80000
Total cost	$200000
The annual sales are now	400*(200+400)=$240000

The annual profit from selling Models B and C is:

240000-200000=$40000

Thus by eliminating the money losing model A profit is reduced from $50000 to $40000.

The reason of course is that the fixed indirect cost is now allocated only to the two models left B and C reducing the profit per unit of each. In reality the above analysis may lead to greater mistakes since direct labor is usually not flexible and eliminating model A does not reduce direct labor cost of the line proportionally.

To overcome the problems of traditional cost accounting systems it is better to base decisions on the analysis of the total cost rather than on unit cost. Another approach is to look at the throughput of each model defined as the difference between its selling price and the costs directly proportional to the volume of production (e.g. the cost of direct material).

Table 4-3. Unit throughput and total throughput per model

	A	B	C
Annual demand	100	200	400
Unit selling price [$]	400	400	400
Direct material cost/unit [$]	100	100	100
Throughput per unit [$]	300	300	300
Throughput per model [$]	30000	60000	120000

The total throughput of 210000 is now compared to the rest of the cost :

Direct labor cost 200*100+100*200+100*400=$80000
Indirect cost $80000
Total operating cost 80000+800000=$160000
Total profit 210000-160000 =$50000

Using the same throughput model if model A is discontinued the total throughput is

60000+120000=$180000

while the operating costs are:

Direct labor cost 200*100+100*400=$60000
Indirect cost $80000
Total operating costs 60000+80000=$140000
Total profit 180000-140000=$40000

Thus the use of throughput as a performance measure is better than using the unit cost computed by traditional cost accounting. As explained earlier the best approach for analyzing major ad hoc decisions like discontinuing a product model is to look at the whole picture i.e. a Profit and Loss statement with and without the proposed change.

The Operations Trainer support cost related decisions by the traditional cost accounting model as well as by the throughput analysis.

4.6. Dynamic aspects -marketing in the Operations Trainer

In the Operations Trainer the market for end products has two segments: large clients and small customers. The clients submit Request for Proposals (RFP) for four types of contracts:

1. One time contracts - the bid is for a single delivery of large quantity of a given end product at given date for a given price.
2. Weekly deliveries - the bid specify the quantities to be supplied weekly, the dates of the first and last delivery and the price. Delivery is required every week on the same day of the week.
3. Monthly deliveries - the bid specify the quantities to be supplied monthly, the dates of the first and last delivery and the price. Delivery is required every month on the same day of the month.
4. Blanket contracts - the bid specify the dates of the first and last delivery and the price. Delivery is required every month on the same day of the month. The exact quantities are announced few days prior to the monthly due date.

The decision to bid for a contract is dynamic in nature as it depends on the current load on the shop floor as well as on the expected load in the future, the proposed price and the duration of the proposed contract.

The demand in the second segment of the market -small customers purchasing few units at a time is affected by two factors - the quoted price and the quoted lead time (QLT) by changing these factors the user can increase or decrease customers demand.

The marketing aspect in the Operations Trainer emphasizes the three dimensions of competition, demand is influenced by the following factors:

- Lead time- the quoted lead-time (QLT). The shorter the QLT the higher the demand. By changing the value of QLT the user can increase or decrease the demand for the corresponding end product.
- Cost - the market is sensitive to the quoted cost of end products, by reducing the quoted cost the user can increase the demand for the corresponding end product while increasing the quoted cost, decreases the corresponding demand. Contracts are awarded based on cost only as each contract has a specific cost that is usually lower than the quoted price per unit.
- Quality - due date performance (DDP) is a measure of quality of the order fulfillment process. A high quality order fulfillment process that delivers on time enhances the reputation of the organization and boosts demand. Poor due date performances hurt the perceived quality and reduce demand. Furthermore, long term contracts are subject to cancellation due to late delivery.

Marketing decisions regarding QLT are strongly related to the inventory policy. When the QLT is too short for a make to order policy to be applied the user can select a make to stock policy. In this case the user dictates the safety stock of each end product and work orders are issued automatically when the on hand inventory level gets below this specified level of the minimum stock.

In addition to automatic work orders the user can issue manual work orders dictating the quantity and type of the end product to be produced. Thus, there are two modes of operation: a policy based mode in which the policy for issuing work orders is the driver for new orders, and a manual -ad hoc mode in which the user decides when to order and how much. Both modes are supported by a Management Information System of the MRP type as explained in Chapter 7.

This system presents updated information on the following:

- Contracts: every contract is listed including due dates quantities and prices.
- Work orders: a Master Production Schedule lists all existing work orders including end products ordered, quantities, due dates and the date the order was issued.
- Sales: last year sales are reported in a graph or a table. This information is available per end product/market combination as well as aggregate information for all the products combined
- Shipments due: a listing of all future shipments is provided along with quantities and promised due dates.

This information is continuously updated and is always timely and accurate.

Problems

1. Discuss the four dimensions of competition: time, cost, flexibility and quality. Explain the trade off between these demotions and use examples of different markets, in which one of the four dimensions is clearly dominant.
2. Explain how competition motivated shorter lead times in the service industry.
3. Select an order fulfillment process of an organization you are familiar with and analyze its lead-time. Explain how is it possible to reduce this lead-time and redesign the process accordingly.
4. Discuss the difference between make to stock and make to order policies in the service industry. Use examples to demonstrate the application of these two policies.
5. Discuss the differences in the format of the MPS in the Operations Trainer and the format of the MPS used in Chapter 4. Which in your opinion is better and why?
6. Analyze the quoted lead-time of the three end products in scenario TS100 in the Operations Trainer. Is it possible to reduce this lead-time and how?
7. Analyze the profit of the three end products in TS100. Using linear programming find the best product mix for this scenario.
8. Analyze the relative attractiveness of the three products in TS100 based on the effect of each product on the Profit and Loss statement for last year.
9. Explain the marketing considerations on which the decision to accept or to reject a new contract in TS100 should be based.
10. Discuss the different assumptions used in the product cost analysis and the product throughput analysis. Explain under what conditions product cost analysis is preferred.

5 PURCHASING AND INVENTORY MANAGEMENT

5.1. The need for outsourcing

Capacity, measured in terms of the maximum possible output level of the order fulfillment process for a given period, is limited in most organizations due to limited resources (facilities, machines, workers, etc.). In the short run, additional capacity can be obtained by increasing the utilization of existing resources (e.g. working overtime or extra shifts). In the long run capacity expansion is possible by adding resources to the order fulfillment process, (e.g. hiring and training new employees, purchasing, installing and operating new machines, and building new facilities).

Frequent changes in the level of resources or in the level of resource utilization may be impossible or may cause instability of the order fulfillment process. Therefore, most organizations use external sources of capacity and only few organizations are completely self supported. Most organizations that deliver goods and provide services purchase some materials, parts and services from outside sources. This process is known as outsourcing.

Outsourcing is a way for an organization to increase its effective capacity without investing substantial capital in facilities, machines, recruiting and training. Outsourcing is useful in the short run as well as in the long run. Furthermore, outsourcing may be a way to enhance the organization's competitive edge in the areas of cost, time, quality, and flexibility by selecting the best sources and by developing long lasting relationships with the selected suppliers.

To ensure that outsourcing enhances the competitiveness of the order fulfillment process, the following points should be carefully evaluated:

1. Which materials, parts and services should be purchased - this is known as the make or buy decision?
2. How to select the suppliers and what kind of relationship to establish with them - this is known as supplier management.
3. What quantities to order, when to place new orders, and how to manage the stocks of purchased goods and materials - this is known as inventory management.

5.3. Suppliers management

For some companies purchases account for approximately 60 percent of the sales dollars (Burt 1989). For many companies purchases account for 30% to 50% of sales dollars. Supplier's management is therefore a key activity in the order fulfillment process. This activity may be subdivided into three subactivities:

- specifications of requirements
- selection of suppliers
- contract management

Specifications of requirements

The purchasing process starts with a definition of the required goods or services. In this definition three issues are addressed:

1. Definition of the required product or service including functional, physical and technical specifications.
2. Definition of the order fulfillment process requirements, including required lead time, size and number of shipments, shipping arrangements and frequency of deliveries.
3. Definition of the quality system the supplier should employ and quality requirements applied to the product or service.

Regarding the first issue, the level of details of the functional and technical definitions may vary considerably. For example, a car manufacturer may approach a supplier of air conditioning systems early on in the design phase of a new car and request him to participate in the design process and design the air-conditioning system for the car's prototype. In this case only functional requirements may be available and the expertise of the supplier is instrumental in transforming these functional requirements into a detailed technical design. In this example the customer decided to use outsourcing for part of the design effort. A related example is a supplier of Printed Circuit Boards (PCB) who receives a complete design for the PCBs required for the control system of the air conditioning unit and manufactures the boards exactly according to the drawings (this process is known as built to print-btp). A third example is a manufacturer of electronic components who supplies standard, off the shelf, components to be installed on the PCBs. In each case, information on the product or service to be delivered is required, but the type of information and its level of detail vary considerably. This type of information helps potential suppliers to assess their technological ability to deliver the required goods or services.

The second issue, definition of the order fulfillment process related requirements helps the potential suppliers assess their operational ability to meet volume and lead time requirements. The two important questions are whether the required capacity will be available for delivery of the specified quantities on time, and whether the required lead time is shorter or longer than the supplier's own order fulfillment process lead time. In which case, (as explained in chapter 4), the supplier may have to adopt a make-to-stock policy to ensure on-time delivery.

The third issue relates to the quality of the product or service as well as the requirements from the supplier's quality system. In the past, the quality of the goods supplied was frequently specified in terms of percent defectives. Special acceptance tests based on random samples were used to evaluate the actual percent defective in each lot shipped by the suppliers and lots with a percent defective higher than specified were rejected. This approach, which is heavily dependent on acceptance tests, assumes that some of the goods shipped are defective. It reduces the organization's ability to compete on cost, quality and lead-time. Testing increases cost and led time, while random sampling does not eliminate the possibility of not detecting a poor quality part and using it, which results in a poor quality product. A different approach is to define the quality of the goods supplied in terms of specifications and the use to be made of them. These specifications are designed to eliminate the need for inspection of coming lots and to enable the trace of quality problems to the source by the supplier in order to eliminate the sources of such problems. Specifying quality requirements for both the goods delivered by the supplier and its order fulfillment process, goods shipped by the supplier can be delivered directly to the point where they are used in the production process, without the need for testing and storage. Thus improving the quality, cost and time performances of the entire order fulfillment process.

Selection of suppliers

When the requirements are clearly defined it is possible to deal with the selection of suppliers. The first step is to decide on the number of suppliers for each purchased part. The traditional western approach is to select several suppliers for each part and to use competition to drive the cost down. Multiple sources is also viewed as a means to reduce dependency on suppliers, thus reducing the risk of losing the source of supply due to strikes or other catastrophes, and reducing the risk of a sudden increase in cost or lead time. An opposite approach is taken by some manufacturing organizations in Japan, where a single supplier is selected for each part and strong ties are established between the supplier and the buyer. The single source supplier is selected after a careful evaluation, which focuses on his ability to enhance the competitiveness of the buyer. Richard Newman (1988) list seven areas of evaluation for potential buyers:

1. Process capability - can the process used by the supplier produce the parts at the specified quality level?
2. Quality assurance - are quality control procedures employed by the supplier adequate in order to maintain consistent quality at the required level?
3. Financial capability - what is the risk associated with doing business with the supplier?
4. Cost structure - what are the actual costs of material, labor and overheads and what are the supplier's profits. High costs and low profit may indicate future problems.

5. Value analysis effort - the supplier's ability to perform value analysis and his past success in performing value analysis indicate how well the supplier understands the product specifications and their relative importance.
6. Production scheduling - The ability of the supplier's production planning and control system to deliver on time, to accommodate changes, and to supply in a short lead time, enhances the competitiveness of the buying organization.
7. Contract performance - The performance measures used to evaluate the supplier's performance

The supplier selection process can also rely on a standard specification of the elements of the quality system. The ISO 9000 series, which comprises of assessment specification for a complete program from design through servicing, is an example of such a standard specification.

Contract management

Once the selection process is completed the focus is on contract management, i.e. the ongoing relationship with the supplier. Pence and Saacke (1988) define three categories of relationships between buyers and suppliers regarding the quality of the order fulfillment process:

1. Inspection - focusing on product inspection to eliminate defects.
2. Prevention - focusing on efforts to build quality into the product and process, the purchaser helps the supplier in developing and maintaining a defect free manufacturing processes.
3. Partnership - a long term relationship between the purchaser and the supplier that is based on teamwork between the two parties in the areas of technology (joint design effort) and quality (a certified supplier that delivers directly to the point of use).

The level of coordination between the order fulfillment processes of the buyer and the supplier can take different forms. The traditional western approach is based on one-time orders, which specify the required quantity, due date, cost and payment terms. This approach is usually linked with inspection - goods are shipped to the receiving area where they are inspected and stored until needed for the order fulfillment process. This approach is based on the theory of inventory management discussed next. A different approach is to sign a long term (say annual) agreement with the supplier specifying the cost per unit, estimated total requirements and the promised lead time. The buyer's planning and control systems issue periodic orders to the supplier. In this form of coordination there is no need to negotiate the price and the terms of every order during the agreement period, and long lasting relationships are established between the two parties. When the buyer and the supplier use ERP systems it is possible to link the systems and transfer the periodical requirements electronically. A third approach is based on the Just-In-Time concept where the supplier makes daily shipments directly to the assembly or the production line. This is possible only if the supplier is certified, no inspection of deliveries is required and the relationship between the two parties is based on a real partnership.

5.4. Inventory management - benefit and cost considerations

Inventory Management is concerned with the policies, decisions, actions, and the monitoring and control of stocks. Stocks of different types are used throughout the order fulfillment process including stocks of:

- Raw materials
- Parts and components
- Work in process
- Finished goods
- Supplies

The use of stocks is motivated by its benefit - it enhances several dimensions of competitiveness while making the management of the order fulfillment process easier as explained next:

- Time based competition - this issue was discussed earlier in Chapter 4. Inventories are used to bridge the gap between the lead-time of the order fulfillment process and the lead-time required by the customers. When the lead-time required by the customers is shorter than the lead-time of the process, inventories can be used to supply the demand during the time difference. It is possible to reduce the lead time by using inventories of finished goods, in process inventories or inventories of raw materials, depending on the policy adopted by the organization (make to stock, make to order, or assemble to order). When the market demands off-the-shelf supply, the only way to compete is to hold inventories of finished goods on the shelf.
- Coping with uncertainty - buffer inventories are used when demand or supply is uncertain. To minimize the effect of a random demand or unreliable suppliers, inventories of finished goods or raw materials are carried. These inventories are readily available to buffer a sudden increase in demand or unexpected delay in supplier's shipments.
- Decoupling activities in the order fulfillment process - inventories are used when two stages in the order fulfillment process are not balanced. For example, when demand is seasonal and management adopts a policy of leveled production, inventories of finished goods are built during the seasons of low demand and consumed during seasons of high demand. Another example is decoupling of operations, when the output of one work center is the only input of another work center. A breakdown in the first work center will cause idle time in the second, as it will not have new

work coming in. By keeping in process inventories between the two work centers the second will not be idle as long as it can work on the in process inventory.

- Cost based competition - carrying inventories to reduce cost is the purpose of most early models for inventory management. For example, when the opportunity arises to purchase raw material at a special, low price, it might be justified to stock some quantities of that raw material for future use. In some cases economy to scale influence the price and by purchasing larger quantities, a lower price per unit can be negotiated. This might justify purchasing in large quantities, building stocks and consuming these stocks over a period of time. Another example is the case of a fixed cost associated with each order. In this case large orders distribute this fixed cost over a large number of units thus reducing the unit cost.
- Technological considerations – Some processes are designed for a batch of a given size. For example, heat treatment may take the same time and cost about the same for a single unit or a batch of several units treated simultaneously. To save on the cost of energy, or if heat treatment does not have sufficient capacity, i.e., it is a bottle neck operation, lot production may be adopted and inventories of treated parts are created. Another example is processes in which the set up time is long relative to the unit processing time. In this case (like the case of a fixed cost of ordering), once a set up is completed larger batches reduce the total time per unit (average set up time per unit plus the unit processing time). If the process is short of capacity and it is a bottleneck, scheduling large batches can increase capacity utilization.

These benefits motivate the use of inventories, but inventories are expensive and create waste. When the cost involved is significant and measurable, a relatively simple trade off between the benefits and costs of inventories can support inventory management decisions. However, in many cases it is difficult to estimate the costs of inventories as explained next.

There are three categories of costs associated with inventories:

1. Capital cost - this is the cost of money tied in the inventory system. The cost of capital invested in the goods stocked, and in the facilities and equipment such as storage space, warehouses, storage racks, material handling equipment, etc.
2. Operating costs - the cost of personnel managing and operating the inventory system, cost of energy needed for air-conditioning, light and for operating material handling equipment, the cost of maintenance of the goods stored and the equipment used for material handling. Overhead costs including tax paid for the facilities etc.
3. Risk related cost - the cost of theft, damage due to fire, water, obsolescence and pilferage, or alternatively the cost of insurance to protect against these risks.

The above costs of inventories may be quite significant and the question whether to carry inventories and how much to carry has been for many years of great interest to researchers and practitioners throughout the world. Early research on the subject focused on the cost of inventories. Today the trend is to take a wider viewpoint and also to consider the disadvantages of inventories whose cost is difficult to estimate. The Just-In-Time approach is based on the assumption that inventories are covering problems in the order fulfillment process and management should expose these problems by reducing inventory levels gradually to a minimum. By exposing and solving the root problems the benefits associated with inventories are attained without paying the inventory-related costs.

The management of raw material inventories, work in process inventories and finished goods inventories at the different stages of the order fulfillment process are a key factor affecting the success of the whole process. Shortages may cause poor due date performances while excess inventories are pure waste. Some inventories are inexpensive while other inventories are too expensive and should be eliminated altogether. Some items have a very long shelf life of several years while other items can be stored for few days at most. Clearly the same policy can not be applied to all the different items in inventory and a method for classifying different items for the purpose of inventory management could greatly support the decision making process. A commonly used method for classifying inventory items is the ABC or Pareto analysis.

Villefredo Pareto studied the distribution of wealth in the 18th century in Milan. He found that twenty percent of the city's families controlled about eighty percent of its wealth. Pareto's findings proved to be more general than the initial purpose of his study. In many populations we find that a small percentage of the population (say 15-25 percent) accounts for a significant portion of a measured factor (say 75-85 percent). This phenomenon is known as the Pareto rule. Figure 5-2 illustrates the Pareto principal.

The Pareto rule is useful in inventory management. In many inventory systems a small group of items account for a significant portion of the capital invested in inventory. The application of Pareto analysis in inventory systems is commonly known as ABC analysis, as inventory items are classified as type A (abut 20% of the most expensive items that account for about 80% of the inventory cost). Type B (the next 30% of the items that account for about 15% of the inventory cost); and type C, (the last 50% of the items that account for about 5% of the inventory cost). Cost, as the significant parameter, may be replaced by other parameters such as the Annual Usage Value (AUV), i.e. the annual demand for an item multiplied by its unit cost.

By classifying inventory items based on cost AUV or other factors, it is possible to develop different inventory management policies for different classes of items. For example, inventories of class A items may be tightly controlled and a great effort taken to order these items just in time when they are needed, while type C items may be ordered in large quantities to achieve economy to scale. An example of a very simple, inventory management model is the two-bin approach for managing type C items. Items in inventory are kept in two bins. The capacity of each bin is enough to cover the maximum possible demand during the lead-time required to replenish the item's inventory. Each time the currently used bin is empty an order for items to fill the bin is placed. This method, widely used before computing power became so inexpensive, is simple, inexpensive to implement and

The inventory is depleting at a constant rate of D units per period, its maximum level is Q and its minimum level is O thus the average inventory is $Q/2$ and the periodic inventory holding cost is $(D/Q)*I_c$.

Thus the total per period cost TC of the system is

$$TC = D*U_c + (D/Q)*O_c + (Q/2)*I_c$$

To find the optimal value of Q take the first derivative of the total cost with respect to Q and set it equal to zero

$$\frac{dTC}{dQ} = -\frac{D*O_c}{Q^2} + \frac{I_c}{2} = 0$$

Solving for Q we get the optimal order size

$$Q^* = \sqrt{\frac{2 \bullet D \bullet O_c}{I_c}}$$

and since the lead-time is constant, each order should be placed when the inventory level is sufficient to cover the demand during the lead-time:

$$R^* = D \bullet LT$$

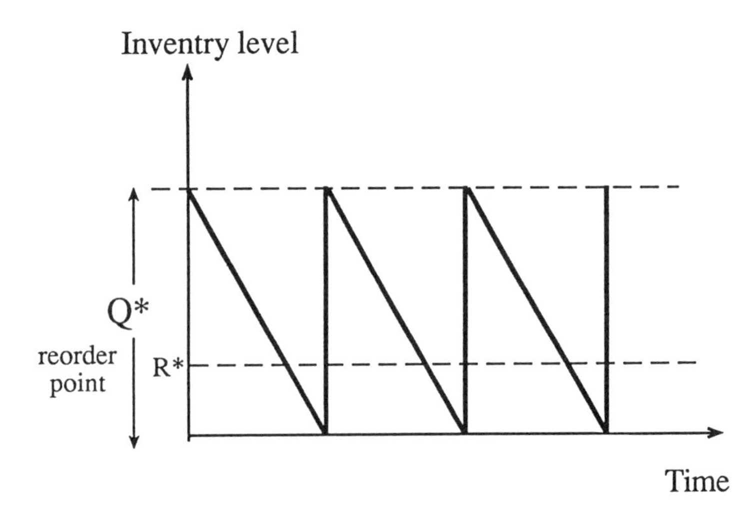

Figure 5.3 Inventory as a function of time in the EOQ model

The various cost components of the EOQ model are illustrated in Figure 5-4. This simple analysis provides an optimal solution to the total cost model under the previously stated assumptions. However, the real question is whether this solution is applicable for real inventory systems. There are some obvious difficulties with the model's assumptions:

- Demand rate is rarely constant and frequently only a forecast of demand is available.
- The costs of real inventory systems vary over time and usually are not exactly known.
- Lead-time is usually a random variable and due to capacity limitations it might be dependent on the size of the order and the current load on the shop floor.
- Interactions exist between different items ordered from the same supplier or delivered by the same order fulfillment process.

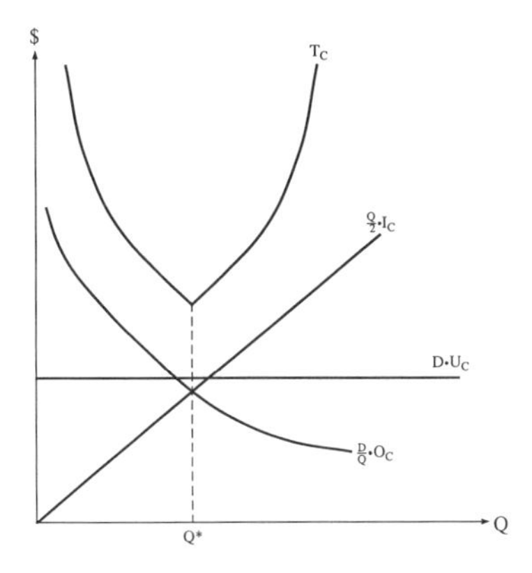

Figure 5.4 The cost structure of the EOQ model

Application of the model to a real inventory system may result in Q* and R* values that are far from optimal.

However, a more difficult problem with the model is the assumption that there are only two decision variables: when to place an order and how much to order. This assumption prevents the user from thinking about the fundamental problem: why carry inventories in the first place. In the EOQ model inventories are carried because of the fixed order cost. By reducing the order cost to zero no inventories will be carried and the optimal solution is to place orders at a rate equal to demand rate. This observation is the basis of the Just in Time approach, i.e. management should try to minimize order cost (or set up cost if production is involved) and to set the order rate equal to the demand rate. As a result no inventories are needed and both the order cost and the inventory carrying costs are eliminated along with other problems created by inventories.

Another problem with the model is the assumption that lead-time is independent of the order size. In reality capacity limits dictate the output rate of the order fulfillment process, and in most real systems lead time will depend not only on the size of the specific order, but also on the other orders competing for the same capacity at the same time. The resource with the most limited capacity in the order fulfillment process will dictate the actual lead-time according to the total load on it. This observation is the basis of the theory of constraints and the Drum Buffer Rope scheduling approach discussed in the next chapter.

The above problems with the EOQ model result from its assumptions. Another source of problems is the lack of integration and the adoption of a static approach. The model tries to isolate the decision on how much to order and when to order an inventory item from the other decisions involved, such as when to order other items; whether to invest in more capacity; whether to invest in set up time reduction efforts; how to market the item, etc. Furthermore the model assumes that the different parameters are constant, while in reality these parameters change over time in an unpredictable way, which is influenced by the decisions taken by management.

An alternative to the EOQ model is the (s,S) model in which the major assumption is that whenever the current inventory level drops below a predetermined value (s), an order is placed to bring the inventory level to a higher predetermined value (S). The advantage of this model over the EOQ model is that it is not based on simplifying assumptions like the basic assumptions of the EOQ model. For example, it is not assumed necessary to place a new order when demand reaches the exact value of the reorder point R. The values of s and S can be calculated under a variety of assumptions including stochastic demand, a cost structure that changes over time, etc. This model sets a framework for a policy in which the parameters can be set by intuition, experience and by applying an integrated approach that considers the whole order fulfillment process simultaneously.

5.6. The dynamics of the order fulfillment process - early studies and the Operations Trainer

The limited ability of quantitative models to capture adequately the integrated, dynamic nature of the order fulfillment process, and to provide sufficient insight into this and other complex business processes motivated the development of alternative teaching and research methodologies. These alternative tools are designed to cope with policy development and analysis in a dynamic environment. Case studies are used for the teaching of complex business situations and simulation models are used to test alternative policies. An interesting example is the Systems Dynamics approach developed by J.W. Forrester (1961). In 1980 Forrester explained the need for Systems Dynamics saying: "The best popular approach to teaching management policy is probably the case - study method. By virtue of its tremendously larger database, the case - study approach has a great advantage over the more mathematical, statistical, and analytical approaches that often dominate operations research, economics and the management sciences. But the case - study method is still hampered by having no adequate way of interpreting its data base into dynamic implications."

An early effort to study the dynamics of the order fulfillment process was published by Forrester in 1961 (Chapter 15 in his book). His production - distribution model deals with the problem of adjusting employment and production to meet market demands and to sustain appropriate inventory levels. In a case study of a real production and distribution system, R. G. Coyle (1978), used Systems Dynamics to study the relationships between production and distribution for an export market. In the study he developed a policy that had the following advantages:

1. A general reduction in stock levels
2. A considerable smoothing of fluctuations in purchase quantity, and hence in production demands on the factory
3. Being very similar, organizationally, to the present system making it very easy to implement
4. Being robust in the sense that it always has much the same performance regardless of errors, or fluctuations in its environment.

These early studies focused on developing and testing policies for the order fulfillment process. However they suffered from two problems common to most System Dynamics models:

- The models did not incorporate ad hoc decision making into the analysis. As Forrester (1980) explains: "A System Dynamics model contains policies that are constant for the duration of a model simulation ...Critics of System Dynamics models lament this lack of recognition of free will... ". The modern simulation languages I-think and Stella partially solve this problem by integrating decision making into the simulation model during a simulation run. Changing some parameters of the simulation model during the run does this.
- The models present an aggregate view and do not capture the details of the system. For example, in a Systems Dynamics model of the order fulfillment process it is impossible to trace a unit of the product or even a specific work order. Since important ad hoc decisions such as expediting are based on detailed information, the aggregate approach of Systems Dynamics do not support detailed analysis of these aspects of the order fulfillment process.

The Operations Trainer is designed to overcome these problems by combining the case study approach with interactive simulation and a MIS system. Quantitative models like the (s,S) inventory model are built into the Trainer and the user can incorporate these models and change their parameters at any point of the simulation run based on very detailed information. Furthermore, the user can always shift to "manual control" in which the decisions are made ad-hoc by the user who overrides the Management Information System.

5.7. The purchasing function in the Operations Trainer

Purchasing in the Operations Trainer is integrated with the rest of the order fulfillment process by the Management Information System that applies MRP logic. This logic, as explained in details in Chapter 7, derive the requirements for purchased materials (independent demand) from the requirements for end products (dependent demand items) taking into account current inventories of finished goods, work in process inventories and inventories of raw materials.

As an alternative the user may apply the (s, S) logic. In this case the user has to set the values for the two parameters, s and S. Both the MRP model and the (s, S) model issue automatic purchase orders. The orders are routed directly to the default supplier for that material. The user selects the default supplier for each material based on relevant information such as the cost per unit, the shipping cost per order and the quoted lead-time or the actual lead-time of the suppliers. Information about the actual lead-time of each supplier is collected automatically and a moving average for actual lead-time during the last six months is calculated.

The user can issue manual purchase orders when desired. Updated, accurate information on the actual level of raw material inventories along with information on open orders due to arrive is continuously available to alert the user of possible raw material shortages that might hamper the order fulfillment process. In the manual mode, the type of material ordered, its quantity and the selected supplier, are input by the user. In some scenarios it is possible to sign a long-term contract with suppliers. In this case a package deal is available -by ordering a monthly or a weekly quantity for a whole year, the purchaser gets a unit price lower than the regular price of that supplier.

The MIS provides information on each material including information on the current purchasing activity and past activity. Detailed purchasing information includes:

- Material information including the current default supplier, unit price, quoted lead time, number of units in stock, current month consumption, last month consumption, the values of the parameters s, S, currently in use and a list of end items where the material is used.
- Last year consumption
- Actual lead time (a moving average over the last six months)
- Purchasing forecast which is based on the Master Production Schedule
- Purchasing performance measures including value of current raw material inventories, the ratio of actual purchase cost per unit of each raw material type to the lowest cost available, number of units not supplied on time to production.
- Open orders that have not arrived yet
- Suppliers information including quoted lead time, unit cost, and shipping cost per order

- Parameters of the (s, S) model
- Safety stock - this parameter is valid for operation under the MRP model, to buffer against uncertainty in supplier's lead-time.

The user controls the purchasing activity by selecting the mode of operation: MRP, (s,S) or manual and by setting the parameters of the selected model. During the simulation the user can use the updated information on the purchasing activities to dynamically change the mode of operation or the input parameters of the current mode. Integration with the rest of the order fulfillment process is maintained by the integrated data base in which information on raw material consumption rate, actual production rate, cash outflow and other important aspects of the process is stored.

Problems

1. Explain the concept of outsourcing using a party as an example. Discuss outsourcing of catering, entertainment, etc. Explain the pros and cons of using outsourcing and the make or buy decisions in this case.
2. Discuss the problems created by limited capacity in a restaurant. Explain how outsourcing can be used in this case.
3. Find an article discussing suppliers selection and suppliers management in the service industry. Explain the differences between outsourcing in manufacturing and outsourcing in the service industry.
4. Explain the differences between subcontracting and purchasing in manufacturing and in the service industry. Discuss the most important factors affecting the decision and the sources of data that support the decision making process.
5. Explain the role of quality and its management in the process of supplier selection.
6. What kinds of inventories are carried out in the service industry in the following sectors: retail , wholesale, health care, education and entertainment . Discuss the benefits and costs of carrying out inventories in each sector and explain how these inventories are used to enhance the competitive edge of the organizations.
7. Explain why utility companies are facing difficulties in using inventories. What kinds of policies are used by utility companies to buffer against uncertainty ?
8. Discuss the cost of inventories in Scenario TS100 based on last year's information.
9. Perform ABC analysis on the inventories carried in Scenario TS100 what are your recommendations ?
10. Calculate the values of s and S for each end product in scenario TS100 .

6 SCHEDULING

6.1. The job shop, implementing priority rules

Time based competition starts with carefully planning the timing of each activity in the order fulfillment process to meet customer's due date requirements. It proceeds with an effort to execute the plan with minimum deviations despite the uncertain and ever changing environment. Scheduling concerns the allocation of limited resources to tasks over time (Pinedo 1995). The driver of all scheduling efforts in the order fulfillment process is the Master Production Schedule (MPS) that sets the timing and quantities of independent demand items and deliveries. The MPS is translated into purchasing orders and production orders. The timely execution of these orders, both in purchasing operations and on the shop floor, is essential to guarantee on time delivery to the customers.

Scheduling on the shop floor concerns the allocation of limited manufacturing resources to the manufacturing operations. A variety of operations scheduling models have been developed and implemented. Each model is based on assumptions regarding the technology used in the process the layout of the shop floor, the objectives of the organization and its constraints, and the environment in which the operations are performed. We start our discussion with the job shop-scheduling problem because the functional layout of the shop floor presents the most general and most difficult scheduling problem. As discussed in Chapter 2, machines in the job shop performing the same function are grouped together. The job shop can process a variety of products, each product requiring a set of operations on the different machines. We define an operation performed on a product as a job. The order of processing, called the routing of each product unit, is assumed known, as well as the processing time of each job (the time to perform an operation on a unit of a product on a specific machine). The set up time required to switch a machine from processing one job to another (a different operation on the same product unit or on another product unit) is also assumed known. Since each part type has its specific operations and specific routing, the number of combinations in which the different parts can be scheduled for processing on the different machines rapidly increases with the number of different parts, different operations and different machines, creating a difficulty to solve this combinatorial problem. The solution of this job shop-scheduling problem is a list of jobs to be performed on each machine, the planned starting time of each job and its planned completion time. A variety of scheduling objectives can be assumed including:

1. On-time completion of each part according to the MPS
2. The completion of all jobs as early as possible
3. Minimization of the time that parts spend in the shop (to minimize in process inventory)
4. Maximization of the utilization of resources by minimizing their idle time
5. Minimization of cost by using less expensive resources (e.g. using regular time and avoiding overtime and extra shifts when these are expensive).

A good schedule optimizes the objective function subject to the applicable constraints for example:

- Each machine can process one job at a time
- Each part can be processed by one machine at a time
- Parts should be processed according to the predetermined routing

The job shop scheduling problem is complicated by the need to set up the machines when switching from one operation to another. When set up time is significant and machine capacity is limited relative to the load on the shop floor, batch processing is frequently adopted. By processing batches of the same product there is no need to set up the machines during the processing of a batch and some set up time is saved.

There are several approaches for developing job shop schedules. The general approach that yields optimal solutions is to model all the different product types, the corresponding routings, the different machines and their capacity and the required quantities and lead times of the different products. This approach is applicable only for very small problems due to the complexity of the models (see for example Pinedo 1995). The simplest approach is to partition the job shop-scheduling problem with n different machines into n problems each with a single machine. The solution of the single machine problem is frequently based on priority rules that dictate the order in which jobs are processed on the machine. In addition, a variety of more complicated scheduling models have been developed for the single machine problem. The success of these scheduling models in solving real life problems has been limited since they are based on a local view (a single machine) of a larger problem.

Priority rules are based on one parameter (simple rules) or some combination of several parameters (complex rules). Since it is difficult to predict which priority rules are the best for a given objective, load, shop configuration and constraints, many simulation studies have been conducted to assess the relative effectiveness of the different rules for a variety of job shops. Simple priority rules include the following:

1. First In First Out (FIFO) - this priority rule is based on the arrival order of jobs. Jobs are processed on each machine in the same order that they arrive to the machine from the previous operation. This rule is very simple to implement.
2. Early Due Date (EDD) - the jobs are scheduled according to their due date. The earliest the due date the higher the priority of the job.

3. Current job - this priority rule is designed to save set up time by processing all jobs that require the same set up in sequence, thus eliminating the need for set up operations. The problem with this rule is that jobs with an early due date may be delayed while other less urgent jobs are being processed.
4. Shortest Processing Time (SPT) - this priority rule tries to minimize the number of jobs waiting in front of a machine, by processing short jobs first. The problem with this rule is that long jobs may be completed late while short jobs are completed earlier than their due date.

In addition to the simple priority rules there is a variety of more complex rules including the following:

1. Critical Ratio (CR) - this rule is based on the difference between the due date and the current date divided by the time required to complete the remaining work. Jobs with smaller value of CR are getting higher priority.
2. Slack Time Remaining (STR) - this rule is based on the difference between the time remaining before the due date and the time required for processing the remaining jobs. The smaller the value of STR the higher the priority of the job.
3. Slack Time Remaining per Operation (STR/OP) - this rule is based on the average slack time per remaining operation calculated as the ratio between STR and the number of remaining operations. Higher priority is assigned to jobs with lower value of STR/OP.

An important visual aid in developing and presenting schedules is the Gantt chart that plots tasks as bars against time. To demonstrate the use of priority rules and the Gantt chart consider the following three jobs as an example:

Three jobs are waiting for processing in front of a machine. The first and the second jobs require another operation on a different machine after the current operation, for the third job the current operation is the last one required. Table 6-1 lists the jobs in the order of their arrival. The remaining operations processing times are also listed.

Table 6-1. Data for scheduling example

Job number	Duration of first operation	Duration of second operation	Due date
I	3	8	20
II	4	6	12
III	5	--	10

Applying the FCFS rule we get the following results:

Table 6-2. The results of the FCFS rule

Job number	Priority	Start time	Finish time	Due date
I	1	0	3	20
II	2	3	7	12
III	3	7	12	10

This schedule is depicted in Figure 6-1.

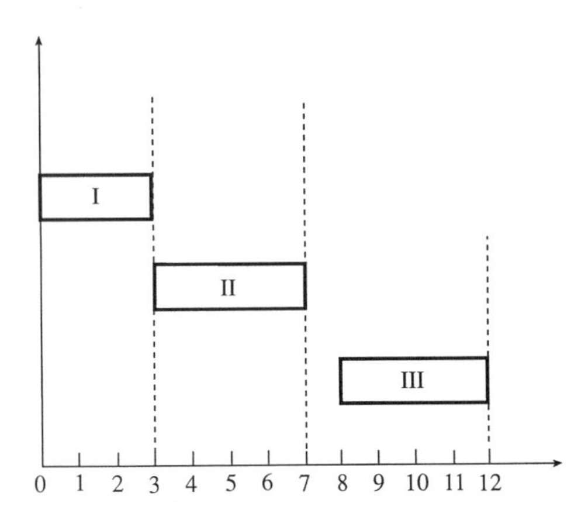

Fig. 6.1 The FCFS Schedule

We see that the third job completes its first operation at time 12, which are two units after its due date. The second job completes its first operation at time 7 and its second operation takes 6 thus it will be completed at the earliest at time 13 while its due date is 12.

Consider the EDD schedule summarized in Table 6-3.

Table 6-3. The EDD schedule

Job number	Priority	Start time	Finish time	Due date
I	3	9	12	20
II	2	5	9	12
III	1	0	5	10

This schedule is depicted in Figure 6-2.

In this case job III completes on time 5 earlier than its due date of 10, job II can start its second operation on time 9 since its duration is 6 it will finish no earlier than time 15 while the due date is time 12. Job I finishes the first operation at time 12, its second operation takes 8 so it can finish on its due date of 20.

Consider the STR schedule summarized in Table 6-4

Table 6-4. The STR schedule

Job number	STR	Priority	Start time	Finish time	Due date
I	20-11=9	3	9	12	20
II	12-10=2	1	0	4	12
III	10-5=5	2	4	9	10

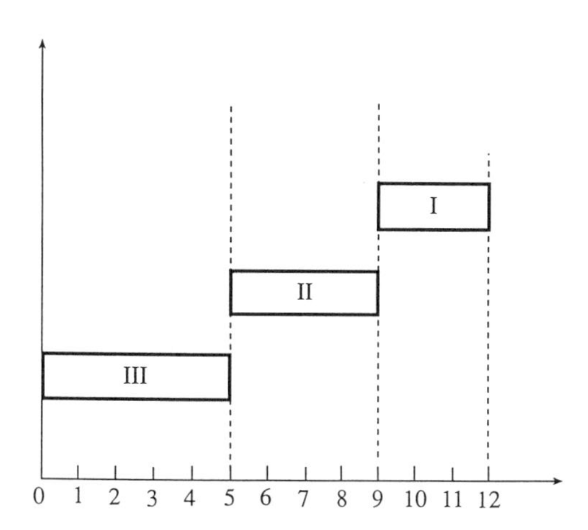

Fig. 6.2 The EDD Schedule

This schedule is depicted in Figure 6-3.

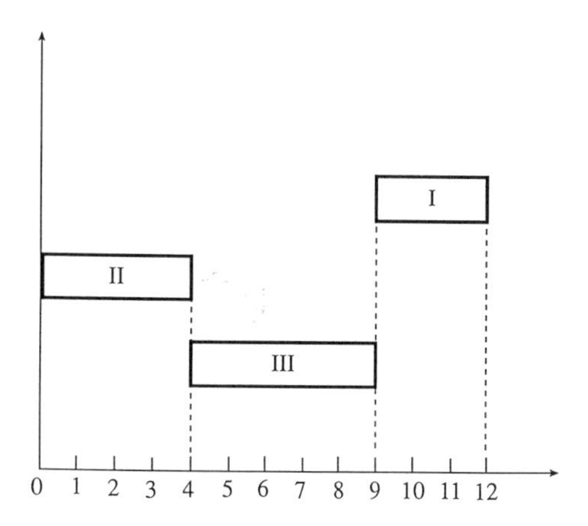

Fig. 6.3 The STR Schedule

In the STR schedule, job II finishes at time 4 since its second operation takes 6 it can finish by the due date of 12. Job I finishes at time 9 which is earlier than its due date of 10. Job III finishes the first operation on time 12, its second operation takes 8 thus it can be finished by its due date of 20.

The above example shows the effect of different priority rules on due date performances of the job shop. By monitoring the situation on the shop floor and selecting the right priority rules for the situation, the management can improve the performances of the delivery schedule. When a job is late it might be necessary to expedite it by assigning high priority to that job. Scheduling jobs manually for processing can do this.

The selection of the most appropriate priority rule, as well as the decision to expedite a job, is based on the shop floor control system. This system provides current information on the jobs waiting for processing in front of each machine either in the form of a list of jobs or as a list of work orders. Another way is to represent the jobs waiting for processing in front of a machine in the form of the total working hours of all the jobs waiting for processing.

Since priority rules are based on the single machine model and ignore the interaction between the machines on the shop floor, the solution developed is at best a good solution for the single machine considered. However, it does not consider the interaction between machines. One undesired result of the interaction is starvation - the case when a machine is idle because all jobs awaiting processing on that machine are stuck as in process inventory in front of preceding machines. Thus the machine is idle while the backlog of jobs that needs its capacity is increasing.

To minimize starvation the in-process inventory levels in front of each machine should be monitored continuously. The control system alerts the order fulfillment team when in-process inventory levels are too low and machine idle time increases or when in-process inventory levels are too high and the lead time increases, due to long waiting times of parts in front of the machines.

An effective tool for scheduling control is the input-output analysis. In this analysis the queue of jobs waiting processing in front of each machine is treated as the level of water in a pool. The input rate of the pool is the stream of jobs coming from other machines while the output rate is determined by the machine production rate (or capacity). When the input rate is higher than available capacity, work in process inventories accumulates. The average lead time of the jobs can be estimated by the ratio between the queue in front of the machine with the highest in-process inventory and the capacity of that machine (both measured in hours). Thus, if a machine operates eight hours a day, five days a week or a total of forty hours a week and the queue in front of it represents eighty hours of work, the average lead time for that machine is two weeks. By controlling the level of work in process in front of each machine or work center, bottlenecks can be identified. The correct schedule of bottlenecks is the key to the lead-time and due date performances of the entire shop. This observation is a corner stone in the Drum Buffer Rope scheduling system as explained later.

6.2. Scheduling the flow shop

In the flow shop, all product types are processed in the same order. The physical arrangement of machines on the shop floor is identical to the order of processing, making material handling easier as all product units simply flow along the common route. As in the job shop, machine starvation is a problem the scheduler should consider. A second problem in the flow shop is blocking. Blocking is the case when a machine is idle because it can not transfer the last product unit it processed to the next machine which is still occupied and the unit stays on the machine and blocks it from processing another product unit. This is typical when the product units are large and in-process inventory is limited, or when the in-process inventory between adjacent machines is limited due to an inventory reduction policy as in the Just-In-Time approach discussed in the next section.

When the capacity of a machine cannot cope with the load, two or more copies of the same machine might be installed in parallel. This configuration is known as the flexible flow shop as there is flexibility in routing the product units between the parallel identical machines.

The simple case of a flow shop with two types of machines can be solved to optimality when the objective is to minimize the makespan (the time from the beginning of the first job on the first machine to the completion of the last job on the second machine). The solution procedure known as Johnson's rule follows:

1. Create two lists, an "available list" of all jobs not scheduled yet and a "schedule list" of jobs already scheduled. The schedule list is composed of two sub lists. The left hand side of the list corresponds to jobs scheduled for early processing while the right hand side of the list corresponds to jobs scheduled for late processing. Thus the first job to be processed is the left most job in the schedule list, while the last job to be performed is the right most job on the schedule list. Initially all the jobs are in the available list while the schedule list is empty.
2. Select the job with the shortest operation time from the available list and delete it from the list, break ties arbitrarily.
3. If the shortest time for processing the selected job is on the first machine, place the job after the last job in the left-hand side of the schedule list. Otherwise, if the shortest time for processing the selected job is on the second machine, place the job before the first job in the right-hand side of the schedule list.
4. If the available list contains only one job go to step 5. Otherwise return to step 2.
5. Assign the last job as the last one in the left hand side sublist and the first one in the right hand side sublist, thus merging the two sublists into a single schedule, than stop at this point, the schedule list contains the optimal schedule.

Example

Consider the four jobs in Table 6-5 and the corresponding processing time on the two machines at the flow shop:

Table 6-5. Data for two-machine flow shop example

Job number	Processing time on the first machine	Processing time on the second machine
I	6	3
II	4	8
III	8	5
IV	2	3

The solution process starts with the available list containing jobs I, II, III, IV and the schedule list is empty. The shortest operation time is that of job IV on the first machine, thus this job is assigned to the schedule list and becomes the first job on the left sublist. Among the remaining three jobs in the available list, the shortest operation is that of job I on the second machine, thus this job is assigned to the schedule list becoming the first job on right sublist. Of the remaining two jobs in the available list the shortest operation is that of job II on the first machine, thus this job is assigned the schedule list becoming second job on the left, i.e. it is scheduled for processing after job IV which is the first on the left. The last job, job III is scheduled after job II and before job I. The final schedule is IV, II, III, I. This schedule is optimal as it minimizes the makespan.

Johnson's rule for the simple case of a two machine flowshop can be extended to a three machine flow shop if the minimum processing time on the first machine is no smaller than the maximum processing time on the second machine, or the minimum processing time on the third machine is no smaller than the maximum processing time on the second machine. This is done by reducing the problem into a two machine flow shop scheduling problem in which the processing time on the new first machine is equal to the total processing time on the first and second machines in the original problem and the processing time on the new second machine is equal to the total processing time on the second and third machine. Applying the Johnson's rule to this new two machine problem yield an optimal solution to the original three machine problem.

Priority rules for the job shop and the Johnson's rule for the flow shop are models based on simplifying assumptions as explained earlier. These models provide some important insight into very simple scheduling problems. Due to simplifying assumptions, direct implementation of their results may lead to poor performances of the order fulfillment process.

A different approach to scheduling is based on the work of Dr. Ohno of Toyota Japan the founder of Just in Time.

6.3. The just in time approach (JIT)

Just in time is a philosophy rooted in the fundamental economic relationship:

$$\text{profit} = \text{revenue} - \text{cost}$$

To increase its profits an organization can increase its revenues by increasing sales (increasing the number of units sold or the price per unit or both), or by reducing cost. Assuming that the volume of sales is inversely related to the selling price of a unit (in many markets, by reducing the selling price it is possible to increase sales) and assuming also that the unit price should exceed the cost per unit to avoid losses, the key to increasing profits is minimizing cost. The focus on cost reduction (or minimization of waste) is the basis of the JIT philosophy and the scheduling techniques implemented in this environment.

There are several, potential forms of waste in the order fulfillment process including:

1. Quality related waste - a defective part represents a waste of material and labor invested in it. A continuous effort to increase quality at the source, i.e. to eliminate the waste associated with the manufacture of defective parts is a key to JIT.
2. Inventory related waste - the money tied up in inventories, as well as the space used for storage and the costs of operating and maintaining the inventory system are all forms of waste that should be eliminated.
3. Waste of space - this form of waste is related to the former, inventory storage is a waste of space as well as inefficient layout of machines that leaves empty space in between machines.
4. Material handling waste - this is the cost of moving material and parts around the factory due to poor layout.

The concentration on waste and its elimination leads to the distinction between value added operations in the order fulfillment process as opposed to non-value added operations. The idea is that every operation in the order fulfillment process should add value to the customer. Operations that do not have a value added should be studied and if possible eliminated. Thus, if material handling on the shop floor does not add any value, a layout that minimizes or eliminates material handling should be investigated. Similarly since inventory does not add any value to the customer it should be eliminated.

Scheduling in the JIT environment is a continuous effort to eliminate waste. Key steps in this effort are:

1. To minimize in process inventories created by batches waiting processing in different work centers, small batch sizes are scheduled for production. When a batch is processed in a work center, and the entire batch is transferred simultaneously from one work center to the next, every unit in the batch waits for the entire batch to be completed. To reduce this waiting time the batch size should be reduced and units should be transferred between work centers as soon as they are finished, i.e. the transfer batch should be smaller than the processing batch. Since smaller batch sizes imply more setups between batches of different products, the key to smaller processing batches is to reduce machine set up time by carefully planning the setup operations and by modifying the machines to enable fast setups that take a few minutes to perform.

2. To further reduce work in process inventories the work in process accumulated between consecutive machines is limited either by the scheduling system or by limiting the space available for in process inventories. Limited in process inventories cause blocking of machines when the succeeding machine fails and starvation when proceeding machines fail or when the output is defective and not suitable for further processing. To minimize blockage and starvation, preventative maintenance efforts and an effort to minimize defects by improving quality at the source are essential in the JIT system.
3. To minimize the inventories of raw materials and supplies, the order cost associated with ordering from suppliers is reduced by purchasing from local suppliers, and by signing long term contracts with suppliers. The simple EOQ model tells us that a smaller order cost implies smaller batch sizes and a smaller inventory level.

The effort to eliminate waste by reducing inventories reduces the buffering and decoupling effects of inventories that were discussed in chapter five. Thus, the order fulfillment process becomes sensitive to scheduling and planning mistakes, quality problems and machine breakdowns. Each of these can cause a delay in the process. To succeed in the elimination of waste while keeping the competitive edge in time based competition, the order fulfillment process in the JIT system is synchronized and controlled by a special system. One way to implement the synchronization system is by a card called Kanban. The Kanban is a card containing information that serves several purposes. It is a production order, a purchase order, and an inventory control device. To understand the Kanban concept, consider two adjacent work centers A and C in a flow shop setting in which jobs move from A to C through a buffer inventory B between the two centers. Two types of Kanbans are used to schedule the two work centers: a production Kanban that functions as a work order for A and a withdrawal Kanban that functions as an order to ship material from B to C. The batch size is fixed (ideally at one).

In this simple two-station example, parts move from A via B to C and Kanban cards move from C to B and from B to A. In each of the three locations there is room for a predetermined, limited number of part canisters and a limited number of Kanbans. Each part canister is accompanied by a Kanban so that the number of canisters is never larger than the number of Kanbans and the in process inventory is thus limited. The Kanbans that move between A and B are called production Kanbans while the Kanbans that move between B and C are called withdrawal Kanbans. The worker at each work center watches the number of Kanbans when the number of withdrawal Kanbans at C reaches a predetermined level, the worker takes the cards to B, removes the production Kanbans from the part canisters in the buffer inventory and attaches the withdrawal Kanbans to the canisters. The production Kanbans at B that are not attached to canisters now serve as an order to the worker at A to produce more units. Canisters with units produced at A are transported to buffer B with the production Kanbans attached to them. The worker at A stops production when there are no more free production Kanbans, thus waiting for the worker at C to withdraw canisters from B and free production Kanbans.

This simple two work centers example demonstrates the principles of a scheduling system that can be extended to any number of stations manufacturing a variety of different parts, as long as each Kanban is related to a specific part, i.e. each Kanban has a part number on it.

This scheduling system is integrated since all work centers are connected to each other by the flow of cards. It is dynamic since the production rate at the centers with excess capacity is controlled by the centers with limited capacity, as in-process inventories are limited to the number of Kanbans in the system. Monden (1983) suggested the following approach for computing the number of Kanbans in the system:

$Y=(DL-W)/A$
Where
Y is the number of Kanbans
D is the expected demand rate (demand in units per unit of time)
L is the lead-time
W is a variable that controls the size of the buffer stock
A is the capacity of a canister

By reducing the value of the control variable W, the in-process inventory is reduced, thus eliminating the waste associated with inventories but also reducing the decoupling and buffering effects of in-process, finished goods and raw material inventories.

Just In Time is a form of Synchronous Production - a system in which the entire order fulfillment process is synchronized, (i.e. works in harmony) to achieve its multidimensional goals (in the dimensions of time, cost, flexibility and quality). The synchronizing device - the Kanban card provides the tight connection between the work centers.

Other forms of Synchronous Production exist as well. As an example consider the Constant Work in Process (CONWIP) system. In its simplest form this system is implemented by a fixed number of containers or pallets on which work in process is transported and processed. The limited number of containers implies a limited number of product units in the system. Unlike the Kanban based system where the limit is on the inventory in each stage of the process, in a CONWIP system the limit is on the total work in process inventory in the system.

6.4. The dynamic shop - expediting and changing priorities

In the JIT approach work orders are generated by actual consumption of parts and material at the succeeding work centers. Since the in-process inventories are limited and production is authorized only when upstream demand consumes some of the limited in-process inventories, quality problems as well as machine break down, cause shortages and idle time at upstream workcenters. When demand is stable over time it is possible to adjust the in-process inventories (or the number of Kanbans) to the demand rate to minimize shortages at upstream workcenters and to ensure a fast order fulfillment process. However, when the demand rate fluctuates over time the adjustment of the in-process inventories may be too slow, causing delays and late deliveries. One way to cope with the problem is to adopt a manufacturing to stock policy, buffering against demand uncertainty by inventories of end products. This approach is unacceptable according to the JIT philosophy as the buffers of finished goods represent waste. Another approach is to freeze the MPS, thus stabilizing demand over the freeze period. This approach reduces the flexibility to react to the changing needs of the market. Since the flexibility of the order fulfillment process provides a competitive edge, freezing of the MPS for the planning horizon may not be a viable alternative. A third approach is to monitor each job as it moves in the shop and change the priorities of work orders according to the changing demand. By watching the situation on the shop floor, comparing the current status of each order to its promised due date and expediting late orders, the management may try to cope with the changing demand and other uncertainties such as machine breakdown, absenteeism and late deliveries by suppliers. However this is easier said than done. Expediting in a real shop is an attempt to solve a very complex combinatorial problem by focusing on the exceptions - those orders that are late. By doing so other orders are delayed and becoming late and the list of late orders tends to increase over time making the problem untraceable. Thus, along with the technology for automatically monitoring the progress of each work order on the shop floor, which is available in today's ERP systems, a systematic approach to analyze the data and to support decision making is needed.

A possible approach is to focus on the problematic workcenters and on the late customer orders simultaneously. Assuming that orders are late due to limited capacity of some work centers. The argument is that management should focus on the work centers with limited capacity and derive the whole schedule for the shop from the schedule of the limited capacity work centers. To apply this approach the monitoring system should alert management to potentially late orders as well as to overloaded resources. By focusing on the overloaded resources, management can utilize the available capacity in the best way, i.e. scheduling these resources in order that potentially late orders are processed first. Furthermore, in the long range, management may want to expand the capacity of resources that are continuously overloaded by outsourcing (using subcontractors), using extra shifts or by acquiring more of these resources. The basics of this approach presented by Goldratt in his book "the Goal " are to identify the "bottleneck" resources and to schedule the whole shop to ensure that these bottlenecks are used effectively and efficiently, i.e. doing what is most important in the best way possible as explained in the next section.

6.5. The drum buffer rope approach (DBR)

The scheduling techniques discussed so far, priority rules for work centers in a job shop, and the Kanban based technique for the JIT environment, represent two extremes in the level of integration between operations in the order fulfillment process. Priority rules are designed to schedule a single work center based on the situation in the waiting line of jobs in front of that work center and based on the center's local objectives. In the JIT environment all work centers are connected to each other by the Kanban system. Production, as well as ordering decisions, are based on global objectives (minimization of waste). Information about the consumption at upper stream work centers and the predetermined work in process between the work centers are used as a basis for scheduling decisions.

In the following discussion a third approach is presented. This approach developed by Goldratt and Fox (1986) is based on a global set of objectives supported by appropriate performance measures and global information about the load on the shop floor. In this approach the schedule of all work centers is driven by the schedule of the critical work centers - those with insufficient capacity.

In Chapter 3 three performance measures suggested by Goldratt and Fox (The Race 1986) were introduced:

- Throughput (T) - The rate at which the system generates money through sales. It is the difference between the dollar sales for the period and the expenses generated by these sales (the expenses that would not occur if the sales were canceled).
- Inventory (I) - The entire amount of the money the system invests in purchasing items the system intends to sell. This is the original purchase cost of raw materials, parts and components stored in the system.
- Operating Expenses (OE) - The entire amount of money the system spends in turning inventory into throughput, which is not directly dependent on the level of sales (i.e. not included in the throughput calculations).

The objective is to maximize T while minimizing I and OE. Every decision regarding the order fulfillment process is judged based on its effect on the three measures. For example, production to stock does not increase T while increasing I and OE. Thus, based on these performance measures production to stock should be avoided (unless it increases T or decreases OE) .

Using these measures, priority on the shop floor is given to firm orders due in the near future that can translate immediately into throughput. An effort is encouraged to minimize all types of inventories in the system (I) and to eliminate costs not directly related to sales (OE). The basic argument that drives the management of the delivery system is that throughput is bounded only if there are one or more constraints that limit sales or production. The constraints can be internal to the system, i.e. a machine, a work center or a worker that have limited capacity. The constraint could also be external to the system such as a shortage of raw material on the market or limited market for the end products. The whole system should focus on the constraint as it is the weakest link in the chain of the order fulfillment process.

To explain the Drum Buffer Rope (DBR) logic, Goldratt and Fox suggested a model of a group of marching soldiers. Assuming that the troop's output is measured by the distance they cover, in-process inventory is measured by the distance between the first soldier in the group and the last one. Naturally the pace of some soldiers is faster than that of other soldiers in the group and it is subject to random variations. Even if, on the average, all the soldiers march at the same pace, when a faster solider slows down for a while due to random variation and then hurries up to close the gap, slower soldiers following him cannot close the gap that fast. Consequently the "in-process inventory" - the distance between the first and the last soldier in the group increases. There are several ways to keep the group together despite the random variations and the differences in pace:

1. To select soldiers with exactly the same deterministic (no random variation) pace.
2. To arrange the soldiers in the group in order of their marching speed. The slower soldiers first and the faster soldiers last. Thus, when random variations occur the faster soldiers at the end can catch up and reduce in process inventory.
3. To use a signal like a drumbeat so that all the soldiers march at the same pace according to the signal. The pace of the signal should be the same as the pace of the slowest soldier in the group.

In this model each soldier corresponds to a work center. The first solution is implemented in automatic production and assembly systems where a group of dedicated machines, all with the same production rate, is used for the mass production of a product.

The second solution is difficult to implement on the shop floor as the order of processing (the routing) is determined by the product processing requirements and usually the routing can not be rearranged according to individual work center's processing speed. In some cases it is possible to implement this solution in a flow shop by selecting faster machines (with higher capacity) for operations performed at the end of the process.

The third solution is the basis of JIT. The drum is the MPS driving the last work center in the process. All other work centers are marching to the same beat as they are connected by the Kanban system.

Goldratt and Fox (1986) suggest a fourth approach. Instead of using the drumbeat as a signal for all the soldiers, use it to signal the first soldier in the group. The slowest soldier determines the drumbeat pace. The analogy is to use the slowest work center (the bottleneck) to signal the pace of the first workcenter in the process - the one that receives raw material from the storage (the gating operation). By linking the bottleneck to the gating operation, release of raw material is synchronized with the capacity of the bottleneck and in process inventories are controlled. This approach is depicted in Figure 6-4.

The problem with this approach is that without in-process inventory, there is no buffer to protect the bottleneck from random failures of preceding machines and to decouple it to avoid blocking and starvation. Thus a predetermined buffer is allowed in front of the bottleneck to protect it. This scheduling approach is known as the Drum Buffer Rope (DBR) where the bottleneck is the "drum" connected by a "rope" to the gating operation and allowing a "buffer" to protect the bottleneck.

The buffer in the DBR system can be managed by input output control. This control system is based on monitoring the in-process inventory buffer in front of the bottleneck. Since the capacity of the bottleneck is known, the lead-time for a new order entering this buffer is the ratio between the size of the buffer measured in terms of work hours and the capacity of the work center. Thus, if for example, the work orders in the buffer require 24 hours on the bottleneck to finish, and the bottleneck operates 8 hours a day, the lead-time is 3 days. Based on the current size of the buffer, a signal to the gating operation to release new work orders is transmitted when the buffer level gets below its predetermined size. Assuming a known capacity of the bottleneck the momentary level of the buffer indicates the lead time of the bottleneck operation.

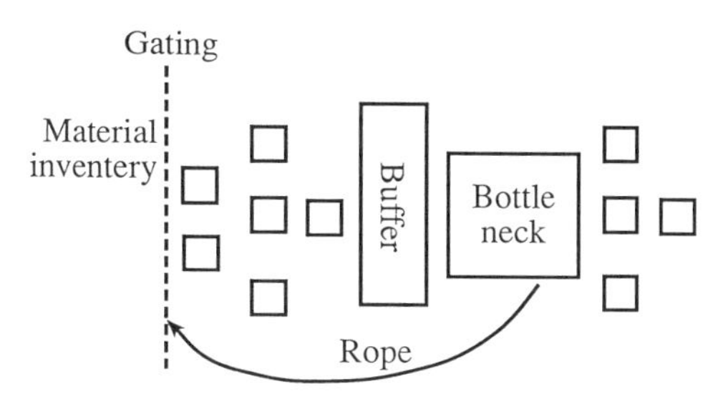

Fig. 6.4 The Drum Buffer Rope Approach

The first step in implementing DBR is to **identify** the system's constraint. An internal constraint that is due to limited capacity is a bottleneck. Comparing the planned load on each work center to its actual capacity can identify bottlenecks. If the planned load is equal to or higher than the available capacity, that resource is a bottleneck. A rough estimate of the load is given by the total processing time required for all jobs scheduled for a work center. The difference between the total processing time and the work center's available time (the scheduled operational time minus expected breakdown time) is the time available for setups. If the minimum required setup time is about the same as the time available for setups the work center is a bottleneck. Otherwise the extra time can be utilized for additional setups (reducing batch size and in-process inventory) or it is an idle time. Since the objective is to minimize I while maximizing T, idle time should not be used for building up in-process inventories, i.e. it is acceptable for a non-bottleneck to idle.

To demonstrate this point consider the daily load for the three work centers example in Table 6-6.

Table 6-6. Data for the example

Job number (number Of units)	I (25 units)		II (50 units)		III (10 units)	
Processing and set up time	Processing time per unit	Setup time	Processing time per unit	Setup time	Processing time Per unit	Setup time
Work center A	2 min	50 min	2 min	100 min	10 min	50 min
Work center B	1 min	50 min	1 min	25 min	20 min	50 min
Work center C	3 min	50 min	2 min	75 min	1 min	40 min

Assuming that each work center is operating 8 hours a day or a total of 480 minutes the processing and set up time for each work center is:

Work center A total processing time:
2*25+2*50+10*10=250 min
minimum set up time
50 +100 +50 =200 min
Thus time left for additional setups 480-250-200=30

Work center B total processing time:

1*25+1*50+20*10=275 min
minimum set up time
50+25+50=125 min
Thus time left for additional setups 480-275-125=80 min
Work center C total processing time:
3*25+2*50+1*10=185 min
minimum set up time
50+75+40=165 min
Thus time left for additional setups 480-185-165=130 min

In this example the bottleneck is work center A with an idle time of 30 min. The capacity of this work center is about the same as its load and it is impossible to split any of the jobs since there is not enough time for an additional setup. The schedule of work center A should utilize its capacity carefully and the work center should be monitored continuously as any loss of capacity may lead to backlogs. Furthermore, work center A should be protected by a buffer inventory to decouple it from problems in its proceeding operations.

In addition to protecting the bottlenecks by buffers, the DBR system tries to move orders as fast as possible through the shop by splitting work orders, i.e. transferring small batches between machines to reduce waiting time. The idea is that the processing batch size is determined by the set up time but the transfer batch size does not have a lower limit and can be reduced to increase T and to reduce I. This is especially important for orders that have been processed by the bottleneck as smaller transfer batches increases throughput without loading the bottleneck. This idea is similar to the use of small lot sizes in the JIT approach.

The DBR is a mixture of JIT and Pareto principles. By applying JIT principles to most of the shop, while buffering the bottlenecks against uncertainty, an improvement in the values of I, T, and OE can be achieved.

6.6. Scheduling in the Operations Trainer

Scheduling in the Operations Trainer has two modes of operations :

- A manual mode applicable to each machine on an individual basis. In this mode the user, according to his preferences, schedules work orders in the MPS waiting processing in front of a machine. The system will prompt the user for new scheduling instructions when needed.
- Automatic mode in which the schedule of each machine is based on one of the following priority rules :

1. FIFO - work orders are scheduled for processing in the order of arrival to the queue in front of the machine.
2. Current Job - to save setups priority is given to work orders that utilize the current setup. Combining work orders that require the same set up saves time. When there are no more work orders that require the current setup in the queue, the FIFO rule is applied to select the next work order for processing.
3. Early Due Date (EDD) - when the current job is finished the queue is searched for the order with the earliest due dates according to the MPS. This work order is selected for processing and when it finishes a new search starts.

In the automatic mode the user can stop a machine from processing the current order, break a setup and manually switch the machine to a different work order.

The generation of new work orders is based on a combination of actual customer orders and a demand forecast. The user controls this process by specifying several parameters:

1. The user specifies a minimum batch size (say equal to the EOQ). If the demand is smaller than the minimum batch size the extra units will be kept as finished goods inventory to fulfill future demand.
2. The user specifies the frequency of planning new work orders. A fixed number of days can be selected as well as a weekly or monthly frequency. In addition the user can generate new work orders manually for any quantity at any time.
3. Special scheduling decisions can be made for contracts. Since contracts represent sizable shipments the user may decide to start processing these orders earlier to avoid overloads and capacity shortages.

To support the scheduling activity information on the current situation on the shop floor is continuously available. The main screen in the production view summarize machine or work center information including:

- Machine name
- The priority rule used for selecting work orders for processing on that machine
- Set up time (in Minutes) for that machine
- The current status of the machine (set up, processing, break or idle)
- The type of product being processed and the operation performed on it
- The current work order processed on the machine
- Number of product units left for processing from the current work order

Additional information is available in the production view: The load on each machine is reported in terms of the total number of processing hours of jobs awaiting in the queue in front of the machine. Also reported is the total number of processing hours on each machine required to process all the work orders released to the shop floor (including work orders that did not reach the machine yet).

Information on the current status of individual work orders is also available. The work orders are represented by the specific routing of the corresponding product and the current location of each product unit from that work order is displayed.

The user can run the factory in automatic mode letting the computer issue work orders and schedule these work orders on the machines based on the priority rules. Alternatively he can issue work orders manually and schedule the machines according to his preferences. Any combination of the two approaches is also possible as the user can monitor the system and intervene at any point of time.

Problems

1. What are the scheduling objectives in a fast food restaurant and in the emergency room of a hospital. Explain the differences and the resulting policies.

2. Discuss the priority rules available in Scenario TS100. Select the most appropriate rule for each machine and explain your decision. Under what conditions these priority rules should be evaluated again?

3. Look at the queue of work orders in front of each machine at the beginning of scenario TS100 and show a Gantt chart resulting from applying the following priority rules to each machine: SPT, CR , STR and Current Job .

4. Explain how Just in Time can be implemented in scenario TS100 and discuss the pros and cons of implementing this policy.

5. Explain how input output analysis is used by department stores for adjusting the number of open check outs.

6. What is the relationship between the capacity of a resource, its cost and the demand rate ? Explain how deviation of actual demand from the forecast may affect the lead time of the order fulfillment process.

7. Using the information on available work in front of each machine at the beginning of Scenario TS100. Explain how the expected lead time of the order fulfillment process can be estimated.

8. What is the bottleneck when scenario TS100 starts. How can you use this information in order to improve the competitiveness of the process?

9. When a new facility is designed, where should the bottleneck be placed. For example, in a new hospital, which resources should have excess capacity and which resources should have limited capacity roughly equal to the forecasted demand ?

10. Explain the relationship between outsourcing decisions and scheduling decisions. If a subcontractor is available with unlimited capacity of all the machines in scenario TS100, under what conditions would you recommend subcontracting?

7 AN INFORMATION SYSTEM FOR OPERATIONS MANAGEMENT - MATERIAL REQUIREMENT PLANNING (MRP)

7.1. The material requirement planning concept

The discussion in Chapter 3 presented the general concepts of Management Information Systems. The discussion on the role of the Marketing, the Purchasing and the Production functions in the order fulfillment process in Chapters 4,5 and 6 presented the application of these MIS concepts to each of the three functional areas involved in the order fulfillment process. A typical MIS for Marketing supports applications like demand forecasting and the management of customer's orders. The MIS for purchasing supports the management of suppliers and the management of purchase orders while a typical MIS for Production provides information like the routing of each part and the resources required for its manufacturing. In reality two or more functional areas share some of the information. For example information on current inventories and open production and purchasing orders is used simultaneously by the Marketing, Production and Purchasing people:

- The Marketing people use this information to determine quantities of products that can be supplied off the shelf, and to decide on when to place new orders for end products.
- The Production people use information on inventories to check that all the material required for a work order will be available when the order is scheduled for production.
- The purchasing people need inventory information to decide on when to order parts and materials and how much to order.

In this case (as in many other cases) the same information should be available to all three functional areas. Furthermore transactions performed by one functional area should update the information used by the other functional areas. Thus if a production order is delayed by the production people due to a machine breakdown, the marketing people should know about it and adjust the promised delivery dates to customers accordingly. The ability of dedicated information systems designed to support one functional area, to support the whole dynamic, integrated order fulfillment process is limited. In addition problems like duplicated data entry (the same data is used in different systems and has to be input several times) motivated efforts to develop integrated Management Information Systems to support the order fulfillment process. Ideally these systems serve not only to support the decision-making processes but also as a mean of communication between the members of the team managing the order fulfillment process. However experience shows that without proper training and education the communication capabilities of these systems are not fully utilized to support group decision-making and teamwork.

Material Requirement Planning (MRP) is an early attempt to develop an integrated MIS for the order fulfillment process. The concept of an integrated system is a significant change from the systems discussed so far:

- By combining information on the demand for products purchased by the customers (known as independent demand items) with information on the structure of these products (subassemblies and components required to make each independent demand item which are known as dependent demand items), the required quantities of dependent demand items do not have to be forecasted - they can be calculated as explained later. Thus the uncertainty associated with forecasting errors is reduced.
- By integrating the inventory management information into the system the requirements for dependent demand items as well as those for independent demand items can be balanced against existing inventories on hand, in process inventories and pipeline inventories so that only the net requirements can be ordered.
- By introducing lead time information for purchased items and for manufactured items manufacturing and purchasing orders can be time phased to ensure delivery exactly when needed.

The original MRP systems (known as MRP I systems) combined the marketing information in the Master Production Schedule with information on current inventory levels and standing manufacturing and purchasing orders, with technological information about the structure of each product and its manufacturing processes. The output included recommendations on how many units of each product, component, parts or raw material to purchase, to manufacture or to assemble and when to issue the production or purchase order.

Over the years new capabilities were added to these systems, including capacity planning modules that reveal capacity shortages, and shop floor control modules that utilize limited capacity efficiently. MRP systems that deal with resource capacities are known as MRP II (or Manufacturing Resource Planning) systems. Recent development in this area produced the ERP or Enterprise Resource Planning systems designed to support the order fulfillment process of an enterprise operating several factories warehouses and an integrated logistic system a complex known as the supply chain.

- Marketing information on actual customer orders and forecasts of future demand. This information is the basis of the MPS, which is translated into the production master plan.
- Manufacturing information on the current load on the shop floor and the ability to supply additional customers orders during each period in the planning horizon. This information is an important input for setting delivery promise dates to customers.
- Purchasing information on supplier's lead-time and the availability of purchased parts and materials in inventory and in the pipelines. This information is important for setting delivery promise dates to customers as it introduces the effect of supplier lead-time into the process.
- Cost information on the cost of manufacturing each independent demand item. When facing with the need to negotiate a price with a customer or to consider discounts to customers this information is very important for the order fulfillment team.

Based on the above information the order fulfillment process management team has to construct and to maintain the MPS. In fact Master Production Scheduling is an excellent example for the need to integrate the different aspects of the order fulfillment process. Although the task of developing and maintaining the MPS may be assigned to a specific person (known as the Master Production Scheduler), this person should be a team player and should understand the relationship between his role and the role of all other players in the order fulfillment process.

The time frame used for the MPS is important. The minimum planning time period known as time buckets specifies the accuracy of the planing process, a time bucket of one week is typical, but shorter or longer time buckets are used as well. The number of time periods that the MPS spans specifies the length of the planing horizon. The minimum length of the planning horizon should be equal to the total time required to purchase raw materials and component parts, to manufacture and assemble the independent demand item with the longest lead time, to provide enough time for the order fulfillment process to supply this item when needed. A longer planning horizon provides better visibility into the future but when it is based on long range forecasts that are subject to high forecasting errors its usefulness might be questionable.

An important part of the order fulfillment process is the Master Production Scheduling modification or change management. Frequent changes in the MPS create changes in the production and purchasing plans that results in nervousness of the system and in low efficiencies, excess inventories and an unstable order fulfillment process. Change management applies different criteria for approving changes in the MPS. By dividing the planning horizon into several time frames (separated by "Time Fences") it is possible to apply appropriate criteria to change requests.

In the short range, changes are rarely approved as the work may already have begun on some orders, materials may already be on the shop floor and machines may be set up for these orders. Some organizations even use the term freezing of the MPS or FIRM MPS, which means that no changes are allowed in the short range.

It is easier to make changes in the MPS if the change affect plans for the far future. In this case, change management may be handled automatically by the computer to insure correct implementation. An intermediate planning period known as the "Slushy period " may also be defined. In this period the Master Scheduler should concentrate on proper validation of proposed changes, evaluate changes and closely monitor their implementation.

The MPS is updated continuously. When the current time period is over, the next period becomes the current one and a new period enters the planning horizon. This process known as rolling planing horizon is designed to keep the MPS current and updated.

In the Operations Trainer the MPS is represented as a list of work orders for independent demand items. For each order the due date and the quantity of the required end product are specified. Key issues to consider are who should manage the MPS and how.

- The Bill of Material -The Bill of Material or BOM is the source of information about the structure of each independent demand item. Through the BOM it is possible to coordinate the requirements for independent demand items with the requirements for subassemblies, components and raw materials as explained later. Assigning an identification number to each component part, raw material, subassembly or product does this. By consistently using the same numbers (known as part number) throughout the MRP system unique identification is possible. Two common BOM formats are the single level BOM and the indented BOM.

The single level BOM shows only immediately required components of each part number. These relationships are known as father - son connections. Information about the components of components etc. is maintained by linking the single level BOMs of different part numbers.

The indented BOM shows all the required components for each independent demand item including components of components and raw materials. To explain the difference consider independent demand item A which is made of two units of X and one unit of Y. While X is a raw material and has no BOM of its own, Y is made of three units of the raw material Z.

The indented BOM of A is as follows:

A
  X-(2 units required)
  Y-(1 unit required)
    Z-(3 units required)

A graphical presentation of the BOM of A is depicted in Figure 7-2

Although the BOM is usually constructed by Engineers its main users are the Production people who "explode the BOM " to calculate the requirements of each subassembly, component part and raw material based on the requirements for independent demand items. Assuming those three units of A are required for week two, then six units of X and three units of Y are required. Furthermore the demand for three units of Y translates into a demand for nine units of Z.

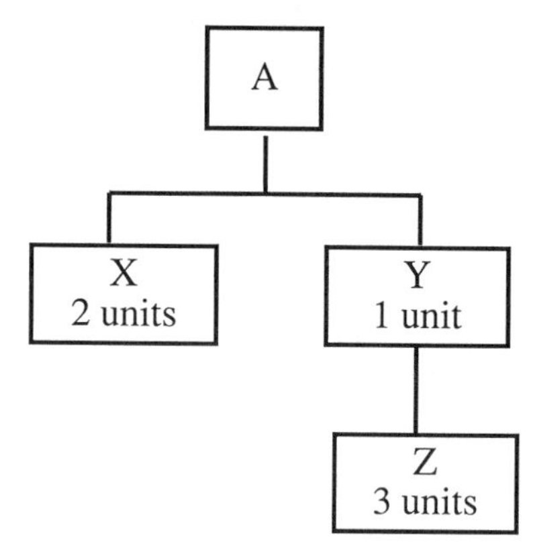

Fig. 7.2 The Bill of Material

In addition to the father-son relationship depicted in the BOM, information on the processing required to manufacture each part number (which is not purchased from outside vendors) is needed as well. This information is stored in the routing file that shows the sequence of operations required for each manufactured part, the machines used to perform these operations and the set up time and processing time per unit. The routing file is the basis of capacity planning and scheduling decisions. In the Operations Trainer the Bill of Material and the routing file combined and the information is presented graphically as illustrated in Figure 7-3

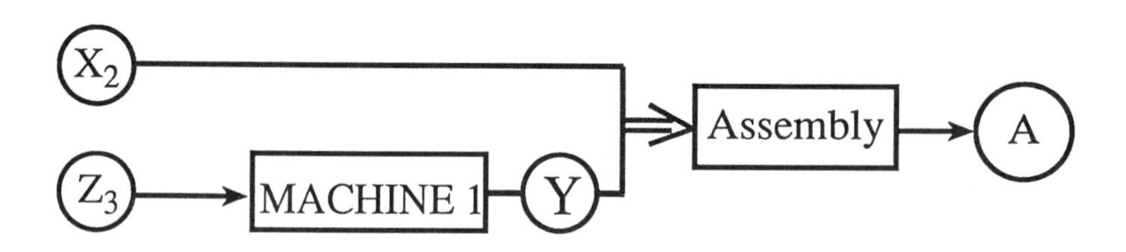

Fig. 7.3 Combined Routing and BOM

Assuming that the purpose of MRP systems is to order exactly what is needed (to minimize inventories), the next question is how many units of each part number are available in inventory. This information is kept in the inventory records.

- The Inventory Records -To function properly, the MRP system compares the gross requirements for each part number to its current inventory. Only if the gross requirements exceed the current inventory an order for that part number should be issued. The current inventory includes inventories in the stock rooms, in process inventories and inventories of parts and material already ordered from suppliers but not yet delivered (known as pipeline inventories). Inventory records contain (for each part number) information on stock on hand, in process and pipeline inventories and the anticipated arrival dates to the factory. The date that in process and pipe line inventories are scheduled to arrive and be ready for use is the basis of the calculation of time phased net requirements as explained later. Inventory information is available to the users of the Operations Trainer: Information on raw material and pipeline inventories is available in the purchasing view, information on finished goods inventories is available in the marketing view while information on work in process inventories is available in the production view. To facilitate an integrated approach, the MRP system combines all this information on inventories in its gross to net computations.

7.3. Gross to net and time phasing - The MRP logic

Material Requirement Planning (MRP) systems are designed to support the material management function in the order fulfillment process . The basic idea is that the same logic can be used for ordering purchased materials or parts, manufactured components and assembled products. The MPS is the source of information on gross requirements for independent demand items. The basic MRP logic makes use of this information as input and translates it into time phased net requirements. The first and second lines in the MRP record of independent demand items are based on the MPS as illustrated in Table 7-2 for the case of product A.

Table 7-2. The basic MRP record for product A

Week	0	1	2	3	4	5	6	7	8	9
Gross requirements			3	7		10		8		6
Scheduled receipts										
Projected available balance										
Planned order release										

Table 7-2 represents the logic of the basic MRP record. The first row indicates the planning time periods (weeks in the example). The second row summarizes the Gross requirements. For independent demand items these requirements are taken from the MPS record. This is the case for product A since it is an independent demand item (see Table 7-1). Gross requirements for dependent demand items are based on the planned order release information in the MRP records of their parents in the BOM as explained later.

The Scheduled receipt information in the third row is related to pipe line and in process inventories. These are inventories for which a work order or a purchase order was issued. The quantity scheduled to be supplied is indicated in this raw in the column that corresponds to the scheduled supply period.

The first entry in the Projected available balance row is the inventory on hand for that part number. This entry indicates the current inventory level. The other entries in this row are estimates of future on hand inventories. These estimates are calculated based on a very simple model, first the projected available balance is calculated:

Projected available balance (t+1) = Projected available balance (t)
+ Scheduled receipt (t+1)
- Gross requirements (t+1)

Based on the calculated projected available balance a decision to issue work orders or purchase orders is made. A simple decision rule known as Lot For Lot (LFL) follows:

As long as the calculated projected available balance is non-negative no order is issued. However when the calculated projected available balance is negative an order is issued and released so that it will arrive (taking the lead-time into account) exactly at the first time period when the calculated projected available balance is negative. The Planned order release field is updated accordingly and the calculated negative projected available balance is increased by the order size.

Assuming that the policy is to keep minimum inventories the order size is set equal to the calculated negative projected available balance .By adjusting the order size in this way, the value of the projected available balance is set to zero. This "Lot For Lot" lot sizing policy is one of several possible policies that will be discussed later.

Assuming that the on hand inventory of A is 5 units and its lead time is 1 week the complete MRP record is depicted in Table 7-3

Table 7-3. The MRP record for part A

Week	0	1	2	3	4	5	6	7	8	9
Gross requirements			3	7		10		8		6
Scheduled receipts										
Projected available balance	5	5	2							
Planned order release			5		10		8		6	

The same MRP record is used for the analysis of requirements for dependent demand items as well .The only difference, as explained earlier, is that the Gross requirements are calculated based on the Planned order releases of the parents in the BOM. For example consider the MRP record of part X. Two units of X are needed for each unit of A and the Gross Requirements for X are 10 units in period 2, 20 units in period 4, 16 units in period 6, and 12 units in period 8. Assuming that there are 35 units of A on hand, 10 units are scheduled to arrive on week 5 and the lead-time of X is 3 weeks, the complete MRP record is depicted in Table 7-4.

Table 7-4. MRP record of part X

Week	0	1	2	3	4	5	6	7	8	9
Gross requirements			10		20		16		12	
Scheduled receipts						10				
Projected available balance	35	35	25	25	5	15	0		0	
Planned order release				1		12				

In a similar way the MRP record is used for each part number. The MPS is used as the source of Gross requirements for independent demand items. Parents Planned order releases are used as the source of Gross requirements for dependent demand items.

This basic MRP logic is modified to accommodate special situations. A common modification is in the lot sizing policy. The lot for lot, lot-sizing logic discussed earlier does not take set up or order cost into account. When these costs are relatively high a minimum batch size is calculated and each time an order is placed its size is set equal to or larger than the minimum batch size. The Economic Order Quantity (EOQ) logic discussed in Chapter 5 is frequently used to calculate this minimum order size. The same logic applies to purchased parts when economy to scale is available i.e. the cost per unit decreases as the order size increases. Other modifications in the lot sizing policy are based on the idea that each order should cover a minimum period of demand. This approach is known as the Periodic Order Quantity (POQ). To demonstrate the minimum order size and the POQ logic consider the MRP record of part A. Assuming that (based on the EOQ model) the minimum batch size is 10 units of A, the MRP record is modified as illustrated in Table 7-5

Table 7-5. EOQ based lot sizing

Week	0	1	2	3	4	5	6	7	8	9
Gross requirements			3	7		10		8		6
Scheduled receipts										
Projected available Balance	5	5	2	5	5	5	5	7	7	1
Planned order release			10		10		10			

Assuming that the POQ model is used and each order should cover the demand of the following four periods the MRP record for part A changes as follows:

Table 7-6. POQ based lot sizing

Week	0	1	2	3	4	5	6	7	8	9
Gross requirements			3	7		10		8		6
Scheduled receipts										
Projected available Balance	5	5	2	10	10			6	6	
Planned order release			15				14			

Unlike the LFL logic both EOQ and POQ generate inventories .The number of orders however is reduced.

Another modification of the basic MRP logic is to buffer against uncertainty. Two types of buffers are commonly used:

Buffer Stock - in this case a minimum inventory level target is set. This value is used in the gross to net computation as follows: The Projected available balance is calculated as explained earlier and an order is placed whenever the Projected available balance is lower than this minimum inventory target. By setting the minimum inventory target to a level that covers the expected fluctuations in the demand for a part number (e.g. fluctuations due to the loss of parts in the assembly process), a buffer against uncertainty is created.

Buffer Lead Time-this method is designed to protect the system from fluctuations in supply lead-time . It is based on increasing the lead-time of a part number by a predetermined amount to protect the system against uncertainty (fluctuations) in actual delivery dates. The result is that on the average shipments arrive earlier than needed and the average inventory in the system increases but the probability of shortages that delay assembly, production or delivery to customers is reduced.

7.4. Capacity considerations

The MRP logic of Gross to Net and Time Phasing derives the requirements for material from the Master Production Schedule. This logic does not check the availability of production resources needed to accomplish the required work on time. This ability known as capacity is a function of the amount of available resources (different machines, material-handling equipment cutting tools and fixtures, storage space and most importantly workers specializing in different operations). Early applications of the MRP logic focused on material requirements, assuming that enough capacity is available in the system to execute all the work orders generated. The problem with this assumption is that in an operation that uses many types of resources the probability that enough capacity of all the different resources will always be available is usually small.

To coordinate resource availability with resource requirements, capacity considerations had to be added to the basic MRP system. The simplest analysis of capacity is based on comparing the resource requirements with resource availability. One way of doing that is known as Rough Cut Capacity Planning (RCCP). Another approach is known as Capacity Requirement Planning (CRP).

Rough cut capacity planning is performed at the MPS level. Its major inputs are the MPS and information about the processing time per unit product on each machine or work center. The logic used for Rough Cut Capacity Planning varies in its complexity and in the accuracy of the capacity requirements forecasts generated. A simple approach to rough cut capacity planning is illustrated in the following four-machine four products example:

Table 7-7 summarizes the per unit processing time in minutes of each of the four products (A, B, C, D) on the four machines. In addition the last column of Table 7-7 presents the MPS information for the next period. The last row in Table 7-7 is a simple estimate of required capacity which is the product of the per unit processing time and the MPS quantity of that product type summed over all product types.

Thus the load on machine1 is calculated as follows:
130*5+110*7+60*3+160*4=2240.

Table 7-7. Data for rough cut capacity planning example

	Machine 1	Machine 2	Machine 3	Machine 4	Total MPS
Product A	5	3	2	5	130
Product B	7	1	4	2	110
Product C	3	8	2	1	60
Product D	4	2	3	3	160
Load (min)	2240	1300	1300	1410	

Assuming that each machine is available five days a week, eight hours a day, the available capacity of each machine is 8*5*60=2400 minutes. Since we did not include set up time in the calculations, machine 1 may be short on capacity. At most, only 2400-2240=160 minutes are available for setups. If we assume that the setup time for this machine is 20 minutes per setup, then a total of 160:20= 8 setups can be scheduled for the period assuming no machine breakdowns and no idle time.

The total setup times as well as the machine breakdown time are difficult to estimate. A simple solution is to forecast these times based on historical data, using a forecasting technique similar to the techniques discussed in Chapter three. Suppose, for example, that using exponential smoothing on the last year's data, the forecasted setup time of machine 1 is 10% of its processing time while machine 1 forecasted breakdown time is 5% of its processing time. Given these forecasts the expected load on machine 1 is 2240*(100%+10%+5%)=2576 minutes. The expected load on machine 1 exceeds its available capacity of 2400 minutes. Management can solve the problem by reducing the load i.e. delaying some of the MPS lots to future periods, or reducing the size of some of the planned lots if the customers agree to partial deliveries. Management can also try to increase the available capacity of machine 1 by operating it on a second shift, by using overtime or by outsourcing-subcontracting some of its load.

Rough Cut Capacity planning is an important tool for the management of the MPS. The decision to accept a new customer order is traditionally based on cost accounting considerations i.e. the comparison between the cost of manufacturing an order and its dollar sale value. The cost of capacity, however is not a simple component in this equation. There is a difference between the cost of an hour of a resource that has idle capacity and the cost of an hour on a resource that is short of capacity. The reason is that for many resources the cost per period is fixed up to a given level of capacity utilization. Additional capacity adds to the cost. A good example is the fixed labor cost of employees on the pay roll. As long as these workers are working up to an agreed number of hours (say eight hours a day) labor cost is fixed. Additional capacity say overtime or extra shifts as well as subcontracting adds to this fixed cost.

Rough cut capacity planning is also the basis for setting priorities in monitoring and controlling the shop floor. Since an hour lost on a resource that has excess capacity is not as significant as an hour lost on heavily loaded resources, it is very important to closely watch the resources that (according to the rough cut capacity planning) have very little slack capacity i.e. to apply tighter control.

The above logic is the basis of the work of Goldratt and Fox (1986) which is an effort to identify the few resources that are short of capacity and to schedule the whole operation in such a way that these resources will be used efficiently and effectively (i.e. they will perform the operations that they do best, in the best way possible) . The assumption is that in a complex system, capacity is never perfectly balanced and therefore only few of the resources will serve as constraints on the production schedule. Goldratt explains this logic in his book "The Goal " and he uses the term "bottlenecks" for these constrained resources. Goldratt's logic is translated into the following steps:

1. Identify the system's constraints (the bottlenecks)- this step can be based on rough estimates of the load on each resource compared to its available capacity (i.e. rough cut capacity planning as discussed earlier in this section).
2. Develop a detailed schedule for the bottlenecks - using simple techniques such as a Gantt Chart as illustrated in Figure 7-4 creates a detailed schedule for each bottleneck.
3. Derive the schedule for all non-bottleneck resources from the schedule developed for the bottlenecks. Since the non-constrained resources have excess capacity the assumption is that it is possible to schedule these resources to support the schedule of the bottlenecks (i.e. to guarantee that the bottleneck will not be idle due to the lack of material to work on).
4. Repeat the process -Since bottlenecks may change over time as the product mix changes or due to changes in resources capacity, it is necessary to identify the system's constraints repetitively every period and to watch the actual utilization of the bottleneck to avoid idle time.

By integrating rough cut capacity planning logic with the above steps a detailed schedule for the heavily loaded resources and supporting schedules for all other resources can be created. A continuos effort to monitor the load on the resources especially the known bottlenecks can alert the team to overload resources that may cause problems in the order fulfillment process.

The use of the MPS as a basis for rough cut capacity planning may lead to substantial errors in estimating the resource requirements. One reason is that some of the resource requirements for any given time period on the MPS may have been fulfilled in earlier periods. This capacity is stored in the form of in process inventories that represent past investment in material and capacity. Furthermore, in any given time period the actual load on resources is generated by the MPS requirements for that period but also by the time phased work orders generated by the MRP for dependent demand items as a result of requirements in future periods.

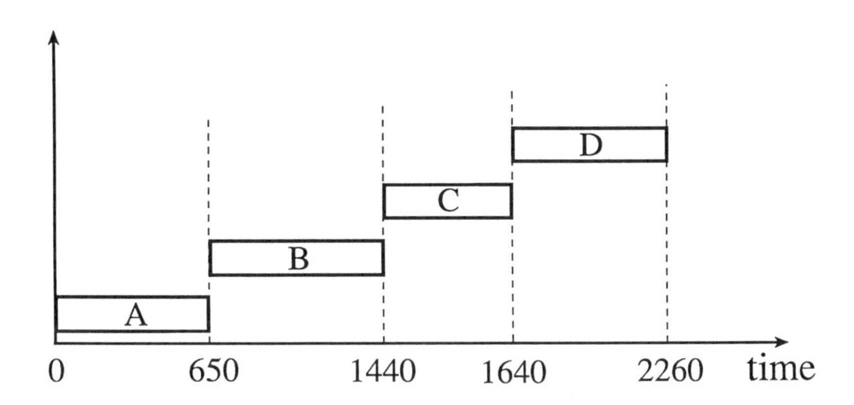

Fig. 7.4 A Gantt Chart for the Example

To overcome these difficulties a more detailed approach for capacity planning is used. This approach known as CRP or Capacity Requirement Planning is based on an effort to treat the requirements for resources by logic similar to the logic used for material planning. This logic is based on the output of the MRP - work orders generated by the Gross to Net and the Time Phasing logic. A resource requirement data base is constructed in which the estimated processing time required for each part number on each machine or work center is stored along with the setup time required to prepare each resource for each type of operation. Based on the orders released by the MRP logic, the required periodical capacity of each resource is calculated. Since these work orders are time phased already, the timing of capacity requirements is more accurate than those of the rough cut capacity planning logic. Furthermore, since the MRP logic that generates work orders takes existing in process inventories into account in its gross to net analysis the capacity requirements are more accurate.

Both CRP and RCCP are tools for testing the feasibility of the plans developed by the MIS. The RCCP tests the feasibility of the MPS while the CRP tests the feasibility of the MRP plans. Both tools are based on estimates of available capacity and forecasts of expected loads and therefore do not provide perfect accuracy. To support the order fulfillment team in its resource management tasks a module called Shop Floor Control (SFC) is added to MRP II system . This module implement logic known as Input- Output analysis. In its simplest form this logic is based on monitoring the actual queue of work orders in front of each work center. By measuring the length of the queue in terms of number of hours required to complete all the work orders waiting for processing in front of a work center and comparing this load to the available capacity of the work center, the time required to complete the current queue can be estimated. This calculated time is a good estimate of the lead-time for that work center. The name Input Output analysis comes from the analogy between the queue in front of a work center and a reservoir. The input rate to the reservoir is analog to the input of work orders to the queue generated by the MRP logic, while the output rate is analog to the rate at which work orders are executed (the capacity of the work center). By controlling the input rate (rate of incoming work orders) the order fulfillment team can control the level of the reservoir (or the in process inventory) of the system. The complete MRP II system consists of the basic MRP I modules plus the RCCP, CRP and SFC modules. The relationship between these modules is depicted in Figure 7-5.

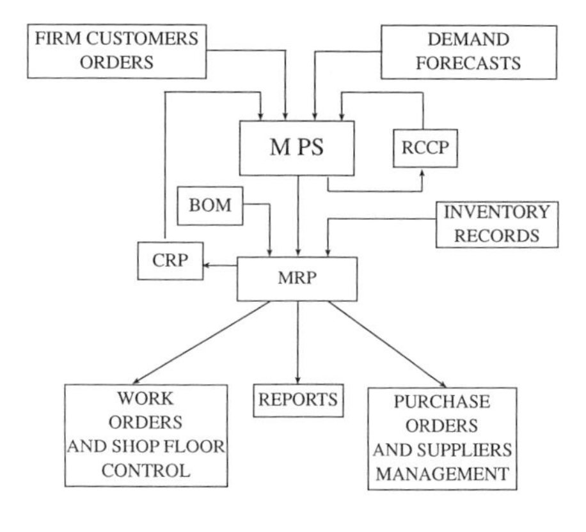

Fig. 7.5 The Modules of MRP II System

The Operations Trainer has a RCCP module built into it as well as simple Input Output logic that shows the number of hours of work awaiting processing in front of each work center. Furthermore the same logic is used to calculate the total number of hours required on each work center by all the work orders released to the floor (even if these orders did not arrive at the work center yet).

Problems

1. Explain how MRP logic can be implemented in a restaurant.
2. Discuss the differences between the EOQ model and the MRP logic. Under what conditions, in your opinion, does the EOQ model outperform the MRP logic?
3. Explain the difference between the MRP concept and the ERP concept.
4. How does an increase in quantity in the MPS gross requirements affect the MRP output? Use the gross to net and time phasing logic in your explanation.
5. How does an increase in the MPS gross requirements affect the available to promise quantity?
6. Explain the difference between a regenerative MRP system and a net change MRP system.
7. Estimate the capacity requirements for the next month for each machine at the beginning of scenario TS100. Compare your estimates with the output of the rough-cut capacity-planning module of the simulator and discuss the differences.
8. Under what conditions do the estimates produced by rough cut capacity planning equal those produced by the Capacity Requirement Planning logic?
9. Suggest a methodology for rough cut capacity planning in a hospital.
10. How would you measure the capacity of a restaurant? An airliner? And a university?

8 MANAGING THE INTEGRATED ORDER FULFILLMENT PROCESS USING ERP: SETTING GOALS, ESTABLISHING PERFORMANCE MEASURES DEVELOPING POLICIES AND TAKING ACTIONS

8.1. The role of management in the integrated order fulfillment process

Management is the art and science of getting things done through other people. Traditionally, organizational structures are used as a framework that supports the management process. Organizational structures establish clear relationships and communication lines between managers and subordinates, within these structures the lines of authority and responsibility are easy to define and maintain.

The selection of an appropriate structure for today's highly competitive environment is not a simple task. In particular, the traditional functional organizational structure, which is based on the principles of division of labor and specialization, may not provide adequate support for the management of the order fulfillment process due to the following problems:

1. Each functional organizational unit tends to focus on its local goals and objectives. The local optimum of each organizational unit does not necessarily lead to a global optimum for the whole organization. For example a common goal for purchasing and material supply is to increase the service level provided, measured as the proportion of requests for material which is supplied on time. One way to achieve this goal is to keep high levels of inventories of purchased items and materials (buffer inventories). Inventories improve the service level but are costly and wasteful. Another way is to assume long lead-times (buffer lead-times) which hurt the time based competition ability of the whole organization. In a similar way marketing is measured on its ability to increase

sales. Thus in a market dominated by time based competition, marketing tends to promise short lead-times to the customers which may cause impossible schedules for the shop floor. Manufacturing is frequently evaluated on its efficiencies, which can be increased by running long batches avoiding set up time. However, long batches may translate into building stocks and thereby reducing the flexibility to respond to changing customer's orders, thus hurting the time based competition ability of the organization. Although the communication within each organizational unit may be easy to maintain, communication between different units may not flow properly. This results from the conflicting goals discussed earlicr, it is also due to differences in the background of people in different functional units, and even due to geographical distances between members of different functional units that are geographically separated.

2. In a functional organization the order fulfillment process does not have a clear owner responsible for its performances from start - getting a customer order, to finish - supplying the order efficiently and on time. Each organizational unit involved is responsible for part of the process, but the whole order fulfillment process does not have a single owner with overall responsibility for setting global goals and the authority to advance these goals.

To overcome these problems of the traditional functional structure in the product development process, a new approach was adopted. This approach, known as Concurrent Engineering is based on a team of experts in the areas of product development, manufacturing and logistics support. Members of the team study the customer needs and evaluate the method of operating the new product. Application of this new approach improved the communication between the participants in the development process and resulted in a shorter development cycle. In a similar way, a team responsible for the whole order fulfillment process can solve the problems of the functional organizational structure. A team that has global responsibility for the whole process, is located together, shares the same information and has a common goal. Team building is a challenging managerial task and it is a process of organizing, staffing, motivating and leading people.

The responsibility of the order fulfillment team is to plan, direct and control the activities of the resources used to perform the order fulfillment process. These activities are related to all three aspects that interact in the process, namely marketing and the interface to the customers, purchasing and the interface with the suppliers and the utilization of resources used for the production of goods and services, i.e. traditional Operations Management.

The basics of marketing, purchasing and operations management discussed in the previous chapters represents the common knowledge that should be shared by all the members of the order fulfillment team. This common knowledge provides the ability to communicate, to set common goals and to develop an integrated plan that can be controlled and directed efficiently by the team. Each team member should specialize in one or more of these aspects, but the common knowledge and the understanding that the whole team shares the responsibility for the entire order fulfillment process is the cornerstone of a successful ERP implementation.

Planning starts with the definition of goals for the whole order fulfillment process and the agreement on the time frame to reach these goals. A combination of long and short-term goals is needed. For example a short-term goal may be to supply a customer's order ahead of it's promised due date when a preferred customer asks for it. An example of a long-term goal is to become a leader in the market and to capture a market share of fifty percent or more within two years. Each goal is a description of a "destination " that the order fulfillment team agrees on and wants to reach within a given time frame.

The development of a plan or a road map that details the steps needed to reach the goal is the next step in the management process. The plan provides each team member with exact information on what he is supposed to do and when. The plan integrates and coordinates the efforts of individuals involved in the process and facilitates teamwork .It serves like the notes that facilitate the integration of the efforts of individual players in the orchestra into a concert. When part of the order fulfillment process is repetitive, the plan can be translated into procedures representing a policy that applies as long as the team agrees to perform the process again and again in the same way.

`A "compass" that provides updated information as to exactly where the team is with respect to its goals is essential. Deviations from the original plan are likely, due to uncertainty such as machine breakdown, late deliveries from suppliers, etc. A set of performance measures corresponding to the goals set by the team and the resulting road map, or the plan developed serves as a compass. By monitoring the value of the performance measures, deviations from the original plans can be traced early on and attended to. Special actions as opposed to long lasting policies are needed when deviations are detected. Detection of deviations and the development of answers to questions, such as when to take an action and which action is most appropriate for a given situation, is part of the management process.

Learning and practicing the art and science of planning, controlling and directing the order fulfillment process, combined with team building is the key to success. Developing a management information system that supports the order fulfillment team is the second key to success. The ERP concept of a single information system that supports all the processes in the organization by providing an integrated data base and a comprehensive model base is therefore essential for the successful implementation. Learning how to use the ERP, how to work in a team, and how to lead in a dynamic, integrated environment is equally important. The Operations Trainer is a tool designed to simulate the order fulfillment process while creating an environment in which the team can practice its managerial skills.

8.2. The hierarchy of goals

In Alice in Wonderland, Alice asks the cat for directions: - should she turn right or left? The cat asks her where does she want to get to and she explains that it does not matter. The cat then replies that in this case she can go either way. In business as in the story, unless we set clear goals (or define where we want to be in the future) we do not know our destination and therefore we can not develop a plan that will lead us there. Frequently the goal for profit-making organizations is to make a profit in the short range and in the long range. To achieve this goal the order fulfillment process must provide customers with the products or services they want, on time, and at a cost low enough to ensure the desired profit given the market price.

This global goal is translated into several lower level goals such as:

- Time related goals :

1. Deliver in a short lead-time, compatible with the lead-time required by customers to guarantee a competitive edge in the market.
2. Deliver on time to the customers, actual delivery date should not deviate from the promised_delivery date.

- Cost related goals:

1. Deliver at the minimum cost possible, a cost low enough to provide the required profit given the competitive market price, i.e. maximizing the throughput of the process.
2. Minimize the operating costs of the process.
3. Minimize inventories to minimize the inventory-related costs.

In this example the upper level goal is translated into lower level goals. Based on these goals a destination as well as a set of performance measures can be defined and used to indicate whether the process is moving toward its specified goals. The next task of the order fulfillment team is to develop a plan that will lead the organization toward its goal.

8.3. Developing a plan - The road map to the goal

To reach its goals the order fulfillment team needs a plan that specifies what should be done by whom and when. A plan that coordinates the efforts of the team members. Coordination is easier in a repetitive environment where the same hierarchy of goals is valid over a long period of time and the same plan may be applied repeatedly to reach these goals. For example, in a make to stock environment, an intermediate goal is to keep the desired inventory level of the make to stock products. The MRP logic described in Chapter 7 can be set to automatically develop a plan of action for this environment by using the Gross to Net and the Time Phasing logic. As long as the goal does not change, the MRP logic is implemented correctly and uncertainty does not intervene, the automatic planning process should yield the desired results, i.e. bring the organization closer to its goal. In this environment the task of management is to set the correct goals (e.g. the desired inventory levels for finished goods) to select the policy, to adjust the MRP logic (e.g. to set the batch sizes and lead-times correctly) and to monitor the performances of the order fulfillment process to detect problems caused by uncertainty and to take care of such problems.

In a non-repetitive environment, this level of automatic planning may not be feasible. In this case each delivery to a customer is a project: a one of a kind endeavor, and new plans have to be developed on a continuous basis. A major task of the order fulfillment team is therefore to identify the repetitive part of the order fulfillment process and to develop appropriate plans that can be implemented continuously. This process of developing the POLICIES that govern the order fulfillment process, results in a generic plan of action that is valid as long as the environment and the goals do not change.

In addition to developing policies the order fulfillment team should take actions to deal with the non-repetitive aspects of the process. The need to take some of these actions may be obvious, for example when a customer calls in and ask for a special delivery (a shorter lead-time, a modified product, etc.) it is clear that a special action is needed.

In other cases a special action is needed because the current policy does not yield the expected results (for example a supplier that did not deliver in the promised due date and cause a delay of a customer order). In these situations management may not be aware of the need to take an action and a special system is needed to bring this need to management's attention. A control system may be used to monitor the performances of the order fulfillment process and to alert management whenever actual performances deviate from the plan and special action may be required. The order fulfillment team has to define what to monitor, how to measure deviations (or irregularities) and when to take an action.

Policies actions and control systems can all be supported by the MIS used by the order fulfillment team. The MRP logic, for example, supports a variety of policies such as lot sizing and the use of buffer inventories. Special actions can be taken by issuing a manual work order, or subcontracting a lot that is usually made in house. Monitoring and control in the MRP environment is provided by logic such as input output analysis that monitors the load on work centers. Easy access to the MRP database provides current information on the status of the order fulfillment process.

The task of the order fulfillment team is to decide which part of the order fulfillment process is repetitive enough to warrant the development of policies to manage it. The team has to develop the policies that will lead the organization toward its goals, to establish an adequate control system that detect situations that should be brought to management attention and to take the correct action when needed.

8.4. Establishing control-identifying problems

In a certain world where perfect updated information about the past, the present and the future is continuously available, the management's task is fairly simple: to establish the order fulfillment team, to develop a proper hierarchy of goals and to select a plan leading to that goal. However, our world is far from being certain - updated, accurate information is not always available: we do not know the future values of many variables that at best can be estimated, and sometimes we do not even know the past or the current values of some variables. Therefore, our plans do not always capture the best way to reach our goals and sometimes our plans turn out to be impossible to implement, due to unforeseen developments. To reduce the effect of uncertainty, management tries to establish systems that identify problems in implementing its plans as early as possible. These monitoring systems are part of a control mechanism that continuously monitor the situation, identify problems preventing the organization from reaching its desired goals and take actions to correct these problems. Continuous control systems are common in many engineering and organizational applications. For example the control of the room's temperature: the thermostat of the system is set to a desired room temperature (the goal) and detects (monitors) deviations between the actual temperature and the desired one. This control system is automatic as no human decision making is involved in the actions taken by the system when a deviation is detected, the air-conditioning system (or the heating system) is automatically activated and changes the room temperature in the desired direction. In this example, deciding on the desired temperature is equivalent to selecting a goal. Setting the thermostat to the desired temperature is equivalent to setting a policy and the control system takes action by turning the air-conditioning system (or the heating system) on and off. Human involvement is thus reduced to selecting goals, setting policies and monitoring the environment to check if there is a need to change the goals, the policies, or both, and to detect any malfunction of the system.

The control of the order fulfillment process is based on the same principles, but more human decision making is required when the process in not repetitive. In this case a computerized monitoring system can detect problems or deviations, but the team managing the process has to decide what to do when a problem is detected, since the decision may be difficult to automate. However, the designer of the order fulfillment process should strive to automate its repetitive parts as much as possible in order that the order fulfillment team can devote its time to the selection of proper goals and the development of policies. The development of monitoring systems is therefore an important task of the team managing the order fulfillment process. A good control system facilitates more automatic decision making as problems can be detected early on and taken care of.

In the design of a control system there are several alternatives. Monitoring the value of the performance measures is one way to go. Assuming that the performance measures are derived from the goals defined for the order fulfillment process, the value of these measures and the trends in these values indicate whether the process is moving in the right direction. The problem with a monitoring system based on the performance measures is that it detects problems fairly late - after influencing the values of the performance measures. If possible it is desired to detect problems before they affect the performance measures. For example if due date performance is an important measure it is preferable to detect problems on the shop floor that might cause a delay in a customer order, or to detect late deliveries from suppliers and to fix these problems before the due date performance level is affected. Thus a system that monitors the load on each work center can send early warning signals when the load exceeds a predetermined value (say two weeks of work). Although the load on work centers may not be a performance measure, its value may be a good indication of the future values of some performance measures (like due date performances).

All three aspects of the order fulfillment process should be monitored continuously: the interface with customers (the demand rate, due date performance etc.) the interface with the suppliers (delivery on time, quality of goods delivered etc.) and the performances of the shop floor (load on work centers, actual lead-times etc.). The monitoring system can operate continuously, i.e. whenever a transaction takes place the values of the relevant parameters are updated, or intermittently, i.e. the parameters are updated periodically. The first approach fits net change MRP systems where each transaction performed is recorded and updates the systems files. The second approach fits regenerative MRP systems where transactions are collected and processed as a batch to update the master files of the MRP system.

The monitoring system can present the value of the parameters selected continuously (as a fuel gage in a car), or it can operate in an on/off alarm mode (like the hot engine light in a car that ignites when the temperature reaches a certain level). The values presented can be momentary values (like the current temperature of the engine or the current load on a work center), or cumulative (like the total distance traveled so far or total sales for the month). The design of the monitoring system is part of the policy development process as the values monitored are those that affect the policy's performances and its ability to reach the desired goal. It is the responsibility of the order fulfillment team to develop a proper monitoring system and to learn how to use it efficiently to support the selected policies.

In addition to the continuous control system, a periodic control should be implemented. Management should take a good look at the process from time to time and evaluate the goals, policies and actions taken in the past in order to improve the process and to adjust it to the ever-changing environment. Periodic control or process design reviews are an opportunity to learn from past mistakes and to continuously improve the process. In the Operations Trainer the users have to specify the duration of each simulation run (a day, a week etc.) the simulation stops after the specified time period giving the users an opportunity to review the current process design , the resulting performances and the actions taken in the last run.

A combination of continuous and periodic control is essential for a successful , flexible order fulfillment process. The design of the control system is an important part of any ERP implementation project.

8.5. Taking actions-solving problems

The monitoring system alerts the order fulfillment team to problems in the process. To solve these problems the team must take proper action. Furthermore the team has to analyze the source of problems and decide whether there is a need to change the current policy in order to avoid the occurrence of similar problems in the future. Thus the management of the order fulfillment process is a continuous effort to design, refine and monitor the process, to look for problems and to solve them. This effort includes the establishment of goals, the development of appropriate performance measures, the establishment of a policy leading toward the goals, the installation of a monitoring system to identify problems and finally taking corrective actions and the modification of goals, policies and the control system when needed.

Problem solving is tricky due to the integrated, dynamic nature of the process and the uncertain environment. Management may take an action related to one aspect of the process and generate other problems and ripple effects over time either in that aspect or in other aspects or both. These system behaviors can be difficult to grasp and even more difficult to predict. Actions are taken under time pressure by learning how to use the capabilities of the ERP system to support group decision making, the team managing the order fulfillment process can improve its problem solving skills.

The first step in solving problems is to recognize that there is a problem and to define it clearly. Problem identification is easier with the support of a monitoring and control system. The sensitivity of the control system should be adjusted so that minor deviations between the plans and actual performances will not cause a "false alarm ". Thus the definition of a significant problem that should be brought to management attention is important. Consider, for example, a control system that monitors the actual delivery of goods from suppliers and compares it to the promised delivery date. A shipment delayed one day may not be a problem if the buffer lead-time used in the system is a couple of days. However, a delay of two or more days should be brought to management attention as it might cause poor due date performances of the whole process.

Once a problem is detected, the next step is to define it in clear terms. The performance measures used to evaluate the process are the basis for problem definition. Thus a possible reduction in due date performance or an increase in the cost of the order fulfillment process are both well-defined problems.

Problem definition is the basis for the next step - problem analysis. This step is aimed at understanding the cause of the problem and its possible effects. It is good practice to use the question "why" repeatedly in the analysis. Thus, in the previous example the problem is a reduction in due date performances. The answer to the question why, is that a supplier delivery is late. Asking why again may lead to the answer that this supplier is always late; i.e. an unreliable supplier (as opposed to the case where something special happened this time that caused the late delivery). If the team asks why do we order from a supplier that is always late the root of the problem may be identified, i.e. that we try to save money.

The third step in the process is to generate alternative solutions to the problem. Two types of solutions are needed: ad-hoc solutions for the current problem and long term solutions to avoid the occurrence of the same problem in the future. The first type of solution leads to an action that solves the current problem, while the second type leads to a change in policy that eliminates this type of problem in the future. Since more than one solution may be generated, the alternative solutions should be evaluated based on their effect on the performance measures and the goals of the order fulfillment process. The alternatives expected to yield the best results are selected and implemented in the process.

This problem solving approach can be summarized as follows:

- Identify the problem
- Define the problem in terms of the process performance measures
- Analyze the problem to find its roots
- Generate alternative long and short range solutions
- Evaluate the solutions with respect to their effect on the performance measures
- Select the best solutions
- Implement the selected solutions

The successful application of this approach by a team requires that the team members understand the details of the process they are dealing with, the goals and performance measures used in managing the process and the monitoring and control system. Furthermore the team members have to understand and practice the concept of group decision making and the tools available to support it within the ERP system.

8.6. Policies control systems and actions in the Operations Trainer

The integrated dynamic management concept is based on management's ability to:

- Understand the whole order fulfillment process and the interactions between its different aspects.
- Establish goals and performance measures.
- Develop adequate policies based on the understanding of the whole process, its goals, its performance measures, its different aspects and the dynamic interactions among these aspects.
- Design a monitoring system and use its signals to control the whole process.
- Develop problem-solving skills as individuals and as a group.
- Implement the policies, controls and solutions to problems.

The Operations Trainer provides a lab-like environment in which a group of managers can develop, practice and test its ability to manage this complex, dynamic, integrated process.

<u>The selection of policies</u>

- Marketing policies

Both make-to-stock, and make-to-order policies are available in the Trainer. In a make-to-stock policy, management can promise a shorter lead-time at the cost of carrying end product inventories. In a make-to-order policy a Just In Time environment is simulated and the inventories of end products are minimized.

Demand is sensitive to the marketing policy, which is controlled by the promised lead-time and the quoted price of the end products. The users control these two parameters and can set them independently for each end product. The market demand increases when the quoted lead-time and the price decrease. If, however actual due date performances are poor, i.e. actual lead-time is longer than the promised lead-time, the market gets the information and demand decreases. As in real life, the exact relationship between the lead-time, the price and demand is not known and the users have to experiment with the marketing policy to get a feeling for the market sensitivity to changes in policy.

Changes in policy take effect from the moment of change but do not affect past events. Thus if the price is changed, the new price will be valid for new orders but the backlog will be sold at the old price.

The levels of safety stocks of finished goods under assemble-to-stock policy are subject to management's decision. Management controls the levels of safety stock and thus the service level to the customers.

- Purchasing policies

The purchasing policies are related to supplier's management and the management of purchased materials.

The management of purchased materials inventory is based on the (s, S) model discussed in Section 5.5. Using this model the ERP continuously reviews the current inventory level defined as stock on hand, plus open orders and compares it to the predetermined reorder level (s). Whenever the current inventory level drops below the reorder level, an order is issued and its size is equal to the difference between the current inventory level and the desired maximum inventory level (S). Management controls this policy by setting the values of (s) and (S). The value of (s) controls the safety stock of purchased material. If no safety stock is desired, (s) is set equal to the demand during lead-time for the purchased material, if higher values of (s) are selected, the difference serves as a safety stock that protects the system against uncertainty. The difference between (s) and (S) controls the frequency of orders or the periodical order cost, as well as the average inventory in the system. Thus (S) is set to minimize the total cost of ordering and carrying inventories.

Management policy regarding suppliers is controlled by the selection of a default supplier to whom new orders are issued automatically by the MIS as long as management does not change its supplier's policy. The selection of suppliers is based on the supplier's lead-time (promised and actual) the cost of purchased materials and the cost of ordering from that supplier. Management can also use statistics collected, regarding actual lead-time to support its supplier's selection policy.

- Production Policies

A dispatching policy is selected for each work center. Based on this policy the worker at the work center decides how to select the next lot for processing from the queue of jobs waiting in front of the work center. Four dispatching rules are available:

i. FIFO - First In First Out: If this policy is selected jobs are processed in the order of arrival to the work center.
ii. CJ - Current Job: This policy is designed to save set-up time. Whenever a job is completed the operator is instructed to look for identical jobs in the queue and process those next to save on set-up time. If there are no more jobs of the same type in the queue, the next job is selected based on the FIFO policy.
iii. EDD - Early Due Date: This policy is based on the due dates of work orders. The operator loads jobs on the machines according to the promised delivery dates to improve the due date performance of the process.
iv. Schedule: This is an ad-hoc approach to scheduling. Management establishes the order of processing at the work center based on any criteria or decision rule. Once the schedule is input the operator will follow it exactly.

The release of raw materials to the shop floor is controlled by the raw material release policy. The simplest form is immediate release, which means that when a work order is generated the required raw material is released to the shop floor. A second policy is to release raw material according to the time phasing logic of the MRP system. In this case material is released only when it is time for the first work center to process it. A third option is manual release. In this case material is released to the shop floor by specific order of management.

Management also controls the policy of using the MRP system. The frequency of planning, i.e. how often work orders are created by merging all the existing marketing demands and issuing a work order is controlled by management. Management controls the batch sizing policy also; the minimum batch size can be calculated using models such as EOQ, POQ, etc.

Control Systems

The Operations Trainer is equipped with a monitoring and control system that alerts management to possible problems in each of the three aspects of the order fulfillment process:

- Marketing control

Management is provided with updated information regarding its due date performances for each end product and an average for all the end products combined. This on line information should be monitored closely as the market is sensitive to due date performance and demand drops sharply if actual due dates do not match the promised due dates. Furthermore, contracts with customers may be canceled due to late deliveries. Canceled contracts reduce the probability of future deals with the same customer.

A "Red Line Control" based on buffer lead-time is used to alert management to potentially late customer shipments. Management defines a buffer lead-time for end products. If on any given day the finished goods inventory of an end product is not sufficient to cover the scheduled shipments for that day and the next few days (that are within the predetermined buffer lead-time), the product icon turn red and in the master production schedule and the potential work orders also turn red. Management can try to solve the problem by expediting the red (late) work orders.

- Production control

The buffer lead-time control system integrates production and marketing. Management can use a red-line policy which means that any work order in the MPS that is late according to the buffer lead-time criteria will have priority in processing, overriding the dispatching rules discussed earlier.

- Purchasing control

A buffer inventory approach is used to monitor the purchasing activity. In addition to the values of the parameters of the (S, s) policy, the user can define a minimum inventory level for each purchased part or material. Whenever the current stock drops below the minimum level a signal is generated and the material icon in the purchasing screen turns red to alert management to the potential problem. Management can set the minimum inventory level so that it will have enough time to receive new material from the supplier with the shortest lead-time (which is usually the more expensive supplier).

- Financial control

Information on the current cash position is continuously available. Management should carefully control the cash flow as deficits' exceeding the available line of credit results in bankruptcy. In addition a profit and loss statement is provided at the beginning of each month to support the financial management of the process.

Actions

By setting policies management directs the order fulfillment process in its steady state i.e. as long as there is no alert from any of the control systems the order fulfillment process should achieve its goals by running according to the predetermined policies. When uncertainty causes deviations from the desired performances the monitoring and control system signals management that it is time to take an action. A variety of actions are available in Marketing, Production and Purchasing:

- Production actions

- Set unit manually- this action is used to switch a work center from the current job processed to another work order. If for example a work order is identified as late by the control system, management may decide to break the current set up on the machine for which the late work order is waiting and start that work order immediately. This expediting action, known as preemption, gives management a complete control over the shop floor.

- Manual scheduling- This action discussed earlier in the context of a policy, switches the task of scheduling from the MIS to management who instruct the selected work center in what order to process the jobs in its input queue.

- Work order generation- Management can generate a work order overriding the MRP system. This can be done by production or marketing by specifying the desired end product and the desired quantity.

- Marketing actions

Marketing, like production, can issue work orders overriding the MRP system. This action can be used if marketing decides to increase temporarily the finished goods stock of any product or to respond to a special customer order.

- Purchasing actions

In addition to the purchase orders generated automatically by the MRP system, management can place special, one-time orders manually. By specifying the raw material to be purchased, its quantity and the preferred supplier, management can override the MRP logic and temporarily increase the stock of raw material, or order from the supplier with the shorter lead-time when shortages are hurting due date performances.

The combination of policy actions and monitoring systems is the key to success in the management of the order fulfillment process. Each scenario requires a different combination of policies, actions and monitoring systems parameters. By practicing different scenarios managers can learn how to analyze a given order fulfillment process, how to develop appropriate policies, how to establish control and how to respond when uncertainty affect the performances.

Practicing group decision-making in a dynamic, uncertain environment is the key to the establishment of a process-oriented organization that deals with the order fulfillment process in an integrated structured manner.

Problems

1. Explain the difference between policies and actions.
2. Suggest a performance measure for each of the four areas in the Operations Trainer and a policy that promotes the proposed performance measure.
3. Suggest a global goal for scenario TS100. Derive performance measures for each area that is related to the proposed goal and suggest a set of policies based on the goal and performance measures.
4. Explain the decision-making process that should be applied in scenario TS100 when a new contract is considered.
5. Develop a linear program that solves for the optimal product mix in scenario TS100 given the load and contracts at the beginning of the simulation.
6. What kind of monitoring and control system is used in a supermarket to support the decision on the number of open checkouts?
7. Explain how should the red level policy be set in the purchasing area.
8. What are the relationships between input output analysis and CRP?
9. Use the problem solving methodology discussed in Chapter 8 to solve the problem of finding the best way to teach operations in a business school setting. Explain each step in the process and the results of your analysis.
10. Develop the ten most important steps in implementing ERP in an organization.

9 TEACHING AND TRAINING INTEGRATED PRODUCTION AND ORDER MANAGEMENT

9.1. Individual learning and organizational learning

The essence of Integrated Production and Order Management is teamwork. A process based organization in which a team is responsible for each process. To ensure a competitive process, each member of the team has to understand the team's task, its objectives, constraints and the performance measures used to evaluate the team. Furthermore, each individual has to learn the concepts of Information Systems, the use of information and the support provided by ERP type systems. This knowledge of individual team members is the basis of coordination and teamwork. It provides the necessary communication channel for group decisions regarding the design of the process and its implementation. In addition to individual learning, team building and team training are major issues in the implementation of IPOM.

The discussion so far focused on the knowledge each individual participating in the order fulfillment process must have. The following discussion focuses on teams, specifically the building teaching and training of the group of people responsible for the design implementation and control of the order fulfillment process.

In the functional organization people are grouped according to the function they perform and are trained to focus on their part of the order fulfillment process. Based on the principles of division of labor, the functional organization tries to divide the process among the people participating in it so that different steps in the process are performed by different people in such a way that these steps are independent (or de coupled) as much as possible. Furthermore, it is assumed that it is enough that each individual performs his part of the process correctly to ensure that the whole process achieves a high level of performances. As a result, in a functional organization each individual is trained to concentrate on his part of the process not having to understand the whole picture. Training individuals to perform their tasks effectively and efficiently is the main effort in functional organizations trying to achieve a desired level of performances.

IPOM is based on an integrated approach to the order fulfillment process (integration as the opposite of de coupling). The assumption is that de coupling is impossible in a competitive, fast changing business environment. In the IPOM environment each member of the team have to understand the whole process including the role of every other team member, as all team members participate in the decision making process. In this environment the team is responsible for the whole process. Consequently teaching IPOM requires a special teaching approach in which individual learning and team learning are combined.

The teaching of IPOM at the individual level starts with the basic concepts of information and its use, following by basic training in marketing, purchasing, scheduling, Material Requirement Planning (MRP) and the concepts of Enterprise Resource Planning (ERP). The teaching at the individual level provides the foundation. The next step is to build a team and to teach it how to work as a team: how to develop policies, how to implement those policies, how to monitor the process and how to deal with problems i.e. to take the correct action when needed. This phase of the training process is similar to the training of a basketball or a football team. Each individual must learn how to play the game and how to excel in it, however a group of good players selected at random does not constitute a good team. Building the team and training it to play together as a team is very important. Each player has to understand his role and to excel in it. Coordination between the players is also an important part of training a basketball or a football team. The high degree of dependency among the players and the dynamic, stochastic environment requires team building and team learning.

9.2. The individual learning curve

The functional organization is based on division of labor and de-coupling of activities. The underlying assumption is that if each individual is an expert in his task the integration of individual expertise by a well-planned organizational structure provides the organization's competitive edge. The expertise of individuals is developed by proper education, training and on the job learning process.

Individual learning can take many forms including the learning of verbal knowledge, intellectual skills, cognitive strategies and attitudes. The learning mechanism can also take many forms including learning by the imitation of other people or learning by repetition of the same process. An early attempt to study the learning by repetition phenomena resulted in the learning curve model of Wright (1936). This model is presented next using the following notation:

- Repetition number
- The time to perform the first repetition
- The time to perform repetition number x
- The learning coefficient

The learning curve model follows:

$T(x) = T(1)x^{-b}$

The shape of the learning curve is illustrated in Figure 9-1

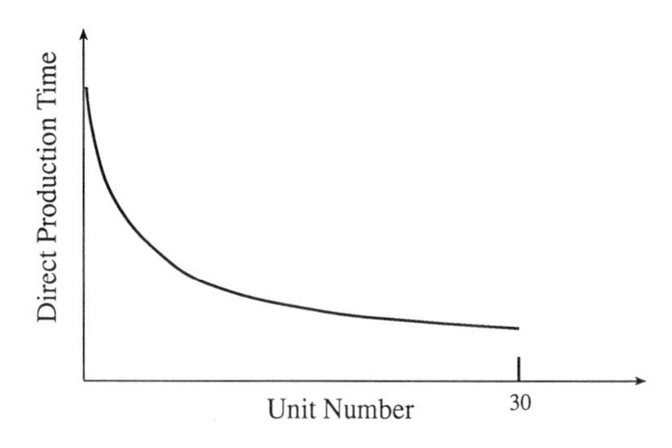

Fig. 9.1 The Learning Curve

In this model the value of the parameter b, known as the learning factor, controls the rate of learning. It is easy to see that whenever the number of repetition doubles the performance time is reduced by the same rate i.e.

$T(2x)/T(x) = 2^{-b}$

Since the introduction of this model by Wright, several other learning models were published. An extensive review of learning curve models is presented in Yelle (1979). The learning curve provides an important motivation for developing tools like the Operations Trainer. The Trainer enables individuals to practice the management of the order fulfillment process repeatedly in a controlled environment and thus moving down the individual learning curve.

The learning curve model is widely used for planning purposes such as estimating the time and cost to complete an order consisting of several units of the same product. It is known that individual learning is enhanced by proper training and by individual incentives.

Group leaning in its simplest form is an aggregation of individual learning. Assuming that the direct labor cost of individuals decreases as experience is gained, it is logical to assume that the direct labor cost of a product or service that represent the aggregate effort of a group of individuals should decrease as a result of individual learning. There are three problems with the extension of the individual learning curve to group learning: first the learning coefficients of individuals in a group are not necessarily the same. Second absenteeism, turnover and job rotation makes the organizational learning curve difficult to predict as each person in the organization may be on a different point on its individual learning curve (i.e. having a different level of experience). In this case those with the lowest experience may be the bottlenecks of the whole process. The third problem relates to synergy in teams. The whole in this case is not necessarily equal to the sum of its components- a good team can perform much better than a group of individuals that divide the work content among them. However a poor team may perform much worse than the prediction of the individual learning curves due to poor coordination between operations performed by different group members. Thus when integration or teamwork is required and tasks are not completely independent and repetitive, the individual learning curve may not be an appropriate model for team learning.

9.3. Team building and the team performance curve

The ability of groups to improve performances and the differences between group learning and individual learning motivated research into this area. One research avenue resulted in the team performance curve model. This model developed by Katzenbach and Smith (1993) relates the type of group and its performance impact as illustrated in Figure 9-2.

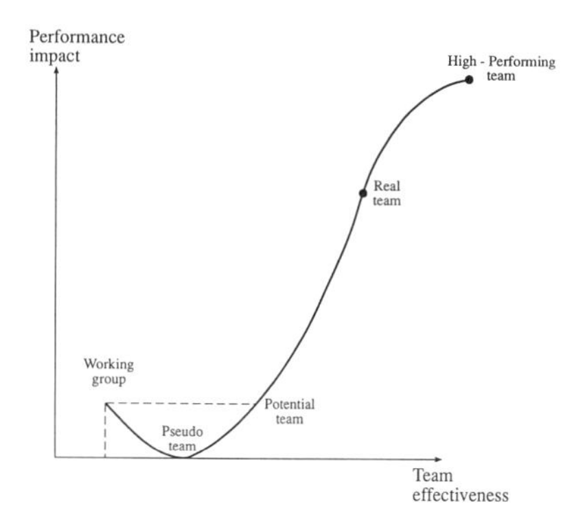

Fig. 9.2 The Team Performance Curve

In this model five types of groups are defined:

1. The working group: the members of the group interact primarily to share information and best practices. They try to make decisions to help each individual perform best within his or her area of responsibility. Working groups rely on "the sum of individual bests" for their performances - their members do not commit to a common purpose and a unified set of goals that leads to mutual accountability. Working groups are common in functional organizations were members from different organizational units cooperate in performing processes that cut across functional lines. Members of effective working groups constructively compete with one another in their pursuit of individual performance targets.

2. The Pseudo - team: In this case members of the group claim that they are a team but they do not really believe in the stated common goals, they do not have a genuine interest in shaping a common purpose and mutual accountability. In some cases the members of the group do not trust each other and can not develop a real commitment to the group. This is the case were the sum of the whole is less then the potential of the individual members.
3. The Potential team: This is a group that recognizes that there is a real need or opportunity for a significant incremental performance and is really trying to achieve it. It has not yet established collective accountability and it requires more clarity about purpose, goals and the common working approach.
4. Real team: This is a group of people who are really committed to a common purpose, goals and a working approach for which the members of the group hold themselves mutually accountable.
5. High - Performance team: This is a group that meets all the conditions of a real team and has members who are deeply committed to each other's success. The high performance team utilize synergy to achieve team performance levels far better that the sum of individual performances.

The Katzenbach and Smith (1993) model explains how important it is to combine individual learning with team building in order to succeed in implementing Integrated Production and Order Management. An attempt to implement IPOM without building a real team or even better a high performance team may result in the deteriorating of performances and a failure.

There are several barriers that a new team must overcome. Thamhain and Wilemon (1979) list the following:

- Unclear objectives - the cornerstone of a real team is a clear definition and agreement on the group's objectives. If the objectives are not clear conflicts, ambiguities and power struggle are highly likely.
- Differing outlooks, priorities and interests - a real team must have a common interest in order to translate the objectives into established common goals and performance measures otherwise a pseudo team will be created.
- Role conflicts - team members should clarify who does what and practice individual roles as well as teamwork to eliminate any ambiguity.
- Personnel selection - the selection of individual members of the team should be based on their commitment to the common goals, professional ability and willingness to work in a team.

Team building is not an easy task and there is no guaranteed "how to" recipe for building teams. Furthermore the team building process is risky as it involves conflicts regarding trust, interdependence and hard work. Katzenbach and Smith (1993) list several hints that can help in the process of team building:

1. Establish urgency and direction - All team members need to believe the team has urgent and worthwhile purposes. The more urgent and meaningful the rationale the more likely it is that a real team will emerge.
2. Select members based on skill and skill potential, not personalities. Teams must have the complementary skills needed to do the job. Three categories of skills are relevant: technical and functional, problem solving and interpersonal. It is important to strike the right balance between members who already possess the needed skill levels versus developing the skill levels after the team gets started.
3. Pay particular attention to first meetings and actions. Initial impressions always mean a great deal. The team leader is introduced in this meeting and the acceptance of his leadership is extremely important.
4. Set some clear rules of behavior. Rules regarding attendance, discussion, confidentiality and contributions are very important and help the team perform its tasks in the early stages. Other rules of conduct are typically developed by the team to help its members achieve the team's goals.
5. Set and seize upon a few immediate performance oriented tasks and goals. By establishing a few challenging yet achievable goals members of the team can start working together right a way a process that forge them together.
6. Challenge the group regularly with fresh facts and information. New information helps the team shape a common purpose and set clearer goals.
7. Spend lots of time together. Creative insights as well as personal bonding are created early on if the team members spend time together.
8. Exploit the power of positive feedback, recognition and reward. Positive or reinforcing feedback helps the process of team building and accelerates it.

The Operations Trainer is designed to support team building. A session with the Trainer starts with a discussion on the need for IPOM and competition in today's environment. The trainees are grouped into teams based on their knowledge and experience in marketing, purchasing, production planning and control and finance. The members of each group manage a simulated company on the Trainer. They are selected so that a good balance of skills exists in the group. The first run is performed in a functional like organization in which each team member is assigned a specific responsibility and a related performance measure. In this run the success of the marketing manager is measured based on the volume of sales, the success of production planning and control is measured by machine utilization and the success of purchasing is measured by the actual cost of raw materials and the level of raw materials inventories. The participants learn that there is a need to coordinate effort in order to succeed i.e. to work as a team. They are instructed to select a team leader and to discuss policy issues thus establishing common goals. Success comes with the experience and the members of the group get positive feedback that enhances the group building process.

Promoting and nurturing individual learning, team building and organizational learning are key elements in the implementation of IPOM. Many organizations take the wrong attitude assuming that the installation of a new ERP type Information System can change their performances and provide the competitive edge in today's global markets. Experience shows that implementation of new Information Systems fails whenever proper training and team building is missing.

9.4. Organizational learning in the IPOM environment

Individual learning and team building are necessary steps in the implementation of IPOM, in most cases these steps are the start of the process of organizational or team learning. Team learning is a process traditionally promoted by organizations facing high levels of uncertainty were processes are composed of highly dependent tasks. As discussed earlier a football or a basketball team is neither relying solely on the ability of each player to play as an individual nor on the common goal of the players to win the game. The players must learn how to coordinate efforts in an uncertain environment were "division of labor" is important but the process is far from being completely repeatable, players are dependent on each other as the situation changes rapidly in an unpredictable way. Thus in many areas in sports team learning is considered a fundamental process that complement individual learning and team building. Members of the team learn how to coordinate their efforts by discussing the role of each individual, by developing "policies" and by playing together in a simulated competition were part of the team represent the opponent and the cost of mistakes is negligible. Similarly, in many armies the training of a soldier starts at the individual level. The next step is to build small groups of soldiers who learn how to work as a team. Next a whole company or a whole battalion are trained together and learn how to coordinate efforts. As in sports, armies are using simulation games were the "enemy" is represented by some of the soldiers or by a computer simulation. The IPOM approach and the Operations Trainer are based on the very same idea. Individual learning is the basis, but it is not sufficient in a highly competitive, dynamic, uncertain environment were the coordinated effort of many individuals is required, tasks are not independent and the process is not completely repeatable. When individual learning reaches a sufficient level and each individual understand the underlying concepts of IPOM and how to use Information Systems, groups are created and assigned the responsibility of managing a simulated company together. Specific goals are defined (e.g. maximum profit maximum Due Date Performance etc.) and the group has to compete with other groups running the same scenario. Pretty soon members of the group develop common purpose and a team spirit and the team is ready for organizational learning.

Organizational or group learning is the process of developing the ability of a group of individuals to improve its performances working as a team to achieve a common goal. Argyris and Schon (1978) define two levels of organizational learning: " Organizational learning involves the detection and correction of error. When the error detected and corrected permits the organization to carry on its present policies or achieve its present objectives, then the error detection and correction process is single loop learning. Double loop learning occurs when an error is detected and corrected in ways that involve the modification of an organization's underlying norms, policies and objectives." This definition fits very well with the concepts of IPOM. The idea that members of the team should develop policies together (policies that can frequently be implemented by the ERP system) and monitor the process to take actions and to change the policies when needed is a fundamental idea in IPOM. However in the same article the authors report that most organizations that do quite well in single loop learning, are having great difficulties in double loop learning. Their general contention is that organizations ordinarily fail to learn on a higher level. Organizations tend to create learning systems that inhibit double loop learning, calling into question their norms, objectives and basic policies.

Based on this study successful implementation of IPOM depends on the ability of the organization to create an environment that encourages single and double loop learning. Detecting and solving problems is not enough- a continuous process of questioning current norms, objectives, basic policies and processes in an effort to eliminate the root of the problem and not only its symptoms is necessary.

Group learning is based on several mechanisms one of which is repetition - the very same mechanism that drives the individual learning curve. Other mechanisms include:

1. The ability to collect and share knowledge so that members of the group can learn from each other's experience.
2. The ability to learn from the experience of other groups or organizations
3. The development of an efficient group decisions making process
4. The ability to share and use information in real time.

A special training environment is needed to teach IPOM an environment that facilitates individual learning, team building, group decision making and the sharing of information. An environment that promotes both single loop and double loop organizational learning i.e. the detection and solution of problems as well as the analysis of current processes and policies and the generation of new policies based on timely information and new knowledge. The Operations Trainer is designed as the teaching environment for IPOM. Its implementation as a training tool is discussed next.

9.5. Teaching IPOM - the Operations Trainer

The Operations Trainer is designed to provide the teaching environment for IPOM. To achieve this goal the Trainer is an integration of three different systems:

1. A simulation system that can simulate a variety of scenarios in great detail. The flow of material is simulated at the level of a single product unit and the shop floor is animated so that the status of each machine is continuously presented.
2. An information system that handles the traditional tasks of transactions processing, management information system and decision support system.

3. An ERP like system composed of an integrated database and a model base. This system performs automatic execution of different policies, it provides information required for decision making, for policy setting and for control.

The Operations Trainer supports individual and organizational learning in three ways:

1. As a tool for teaching individuals the concepts of information and its use, along with the basics of scheduling, marketing, and purchasing, Material Requirement Planning and ERP systems.
2. As a tool for practicing traditional Operations Management by assigning a group of students or managers to manage the simulated plant. Each participant has a definite role: marketing, purchasing, finance and operations and each participant is using the data and models of its functional area only. In this case the performances of each individual are measured by a function specific performance measure such as total sales for marketing, cost of raw materials for purchasing, machine utilization for production and cash flow for finance.
3. As a tool for teaching IPOM and the ERP concept by assigning a group of managers or students to manage a simulated plant were the whole group serve as a team responsible for the integrated process and sharing the same information. In this case the whole team is responsible for a set of integrated performance measures such as net profit and due date performances.

In the first and second steps the individual learning curve applies. Users of the trainer learn how to find the information they need and how to develop policies that promote the functional areas for which they are responsible. They learn that there is a clear trade-off between the goals of the different functions. For example in order to increase sales marketing likes to promise a short lead-time which is hurting due date performances. Purchasing tends to purchase from the lowest cost supplier, which in most scenarios is unreliable and has a relatively long lead-time. Production likes a make to stock policy that buffers against uncertainty but hurt the cash flow. Such conflicts lead the members of the group to the conclusion that a common goal and a balanced set of performance measures are required - the group becomes ready to transform into a team.

The team building process is probably the most important step in implementing IPOM. Teaching people how to work to achieve a common goal, how to share information and how to implement group decision making that results in a consensus is unique to the IPOM environment and to the teaching by the Operations Trainer.

The introduction of new information systems is a unique opportunity to implement a change. The combination of a new ERP system and IPOM- the new approach to the management of the order fulfillment process presents an integrated process designed to gain a competitive advantage in today's markets.

Problems

1. Discuss two mechanisms for individual learning.
2. Explain the effect of repetition on quality. Is this a learning effect?
3. Explain the difference between on the job training and other forms of training discuss the pros and cons of each form.
4. Compare between the teaching methodology in elementary schools and the methodology used in universities. Discusses the applicability of each to the training of new employees in your organization.
5. Explain the difference between a group of individuals and a team. Give an example of each and discuss the transition process between the two.
6. Develop a training program for a new individual hired as a waiter in a restaurant. Is there a need for team building in this case?
7. Explain the role of simulation in team building. Discuss two examples.
8. What is the difference between individual learning and organizational learning in a class learning how to use e-mail?
9. Discuss the role of computers in individual and organizational learning.
10. "Learning is a never ending process" discuss the implications for individuals and organizations.

REFERENCES

Argyris C, Schon D. Organizational Learning. London: Addison Wesley 1978.

Blackburn JD. Time -Based Competition. Homewood, Illinois: Business One Irwin, 1991.

Bowersox JD, Closs JD. *Logistical Management, The Integrated Supply Chain Process*, New York, NY: McGraw Hill. 1996.

Box GEP, Jenkins GM. *Time Series Analysis: Forecasting and Control.* 2nd ed. Oakland, California: Holden-Day, 1976.

Burt N. David. Managing product quality through strategic purchasing. Sloan Management Review, Spring 1989; 39-48.

Chase DB, Aquilano JN. *Production and Operations Management: Manufacturing and Services*, Seventh Edition, Chicago: Irwin, 1995

Coyle RG. *Management System Dynamics*, Chichester :Wiley,1978.

Deming EV. *Out of the Crisis.* Cambridge, Mass: MIT Center for Advanced Engineering Study, 1982.

Emmons H. Flowers A. Dale, Khot M. Chamndrashekhar and Mathue, Kamlesh, Storm. *Quantitative Modeling for Decision Support*, Englewood Cliffs, NY: Prentice Hall, 1992.

Forrester JW. *Industrial Dynamics*, Cambridge, Mass :MIT Press, 1961.

Forrester JW. *Principles of Systems,* Cambridge, Mass: MIT Press, 1968.

Forrester JW. "Systems Dynamics -Future Opportunities". In *Studies in Management Sciences*,. A.. Legasto, J.W Forrester, and, J.M. Lyneis. eds. Vol 14 - Systems Dynamics, North Holland,1980.

Gleans J. Ulric, Oram, E. Allan, Wiggins P. William. *Accounting Information Systems,* 2nd Ed. Cincinnati, Ohio: South - Western Publishing, 1993.

Garvin A. David, Competing on the eight dimensions of quality. Harvard Business Review, November - December, 1987

Goldrattt EM, Cox J. *The Goal.* NY: North River Press, 1984.

Goldrattt EM, Fox RE. *The Race.* NY: North River Press, 1986.

Hamilton E, Flowers AD, Chandrashekhar M.K, Kamlesh M. *STORM Personal Version* 3.0, Englewood Cliffs, New Jersey: Prentice Hall, 1992.

Hammer, Michael, Champy, James. *Reengineering the Corporation, A Manifesto for Business Revolution,* New York: Harper Collins, 1993.

Katzenbach, R. Jon, Smith K. Douglas. *The Wisdom of Teams*, Boston, Massachusetts: Harvard Business School Press, 1993.

Laudon C. Kenneth, Laudon P. Jane. *Essentials of Management Information Systems :Organization and Technology*, New Jersey: Prentice Hall, 1995.

Monden Y, *Toyota Production System*. NC: IIE Press, 1983.

Nevis J. James, Whitney E. Daniel, *Concurrent Design of Products & Processes*, New York. McGraw Hill, 1989.

Newman G. Richard, Single source qualification. Journal of Purchasing and Materials Management,Summer 1988; 10-17.

Ohno T. *Toyota Production System : Beyond Large-Scale Production* Cambridge, Mass . Productivity Press, 1988.

Pence L. John, Saacke, P. A survey of companies that demand supply quality. Milwaukee: ASQC Quality Congress Transactions, 1988;.715-722.

Pinedo M. *Scheduling Theory, Algorithms and Systems*. Englewoods Cliffs, New Jersey: Prentice Hall, 1995.

Porteus EL, Investing in reduced set up in EOQ model. Management Science 1985; 31: 998-1010.

Shingo, Shiego A. *Revolution in Manufacturing: The SMED System*. Tokyo, Japan Management Association, 1983.

Shingo, Shiego, Zero *Quality Control: Source Inspection and the Poka-Yoke System*. Stamford, CT: Productivity Press, 1986.

Shtub A, Bard J, Globerson, S. *Project Management Engineering, Technology and Implementation*. Englewood Cliffs, New Jersey: Prentice Hall, 1994.

Silver EA, Pyke DF, Peterson R. *Inventory Management and Production Planning and Scheduling*. third edition New York, Wiley, 1998

Smith Adam. *An Inquiry into the Nature and Causes of the Wealth of Nations*, A Strahan & T . Cadell, London . 1776.

Thamhain HJ, Wilemon DL. Team building in project management. Proceeding of the 21st Annual Symposium of the Project Management Institute, October 1979.

Wright TP. Factors affecting the cost of airplanes. Journal of Aeronautical Sciences, February 1936.

Yelle LE. The learning curves: historical review and comprehensive survey. Decision Sciences, April, 1979; Vol. 10 No. 2: 302-328.

APPENDIX A

A.1 The Operations Trainer (OT)

The OT is a teaching aid that integrates the case study and the modeling approaches into a dynamic teaching environment. It combines a simulation of a specific order fulfillment process (a scenario or a case study) with an ERP like information system. Different scenarios (or case studies) can be taught using the OT.

The OT provides real time, detailed information on the flow of material. It is possible to track a single work order or a product unit as it flows through the order fulfillment process.

A Management Information System (MIS) and a Decision Support System (DSS) combining a database and a model base support the users of the OT who interactively manage the scenario. It is users that set the policies that govern "automatic decision making". The users can override this automatic decision making mode by switching to one time (ad hoc) actions. The OT supports four functional areas: Production, Marketing, Purchasing and Finance. Each area has a set of policies to choose from. For example production can select the most appropriate queuing policy for orders waiting processing in front of each machine. Policies such as First In First Out (FIFO) or Early Due Date (EDD) are available along with a manual scheduling option. Higher level or, integrated policies involving two or more functional areas are available as well. For example production and marketing can select a make to stock or a make to order policy. The users have to analyze the scenario (the case study) and to decide on the most appropriate policies for the scenario. During the simulation run the users get real time information regarding the current state of materials flowing through the order fulfillment process. They also get the current value of several performance measures such as Due Date Performance (DDP) measured as the percentage of product units supplied on time and the profit they make each period. Based on this information they can change policies or take one time (ad hoc) actions. A simple yet effective model base supports the users in their decision making. A variety of well-known models are available. For example-forecasting models (used to forecast demand for end products, to forecast future raw materials consumption and to forecast the required capacity at each work center) are available in the model base and can be applied at any time to the data generated by the simulation.

In addition to planning the process the users learn how to exercise control under uncertainty. Control limits for different parameters such as raw material inventories and finished good inventories can be set and the system alerts the users when actual values of the parameters (e.g. inventories) are outside the predetermined control limits.

The OT is an interactive teaching tool. It demonstrates dynamically the details of the order fulfillment process including the flow of material, the flow of information, and the decision-making processes. The OT supports the proposed methodology for individual learning, for team building and for team training.

A.2 Scenarios in the Operations Trainer

The Operations Trainer can simulate a large variety of scenarios. Each scenario is a dynamic case study characterized by the specific order fulfillment of the simulated organization. The case study describes the organization, its markets, its financial situation, its products and manufacturing facilities, etc. The commercial version of the Trainer is equipped with a scenario generator with which the user can generate the scenarios required for a specific workshop. The student version that accompanies the book comes with a limited number of scenarios representing a flow shop, a job shop and an assembly operation. The three scenarios are sharing many similar characteristics and the student can study the commonalties and the differences between these three situations by comparing the three scenarios.

A.2.1 The flow shop (TS100)

This is the basic scenario:
A small manufacturer of parts for the automotive industry started operations in January 1995. This manufacturer manufactures three products; A1, B1, and C1 all of them are parts of the same subassembly. The three products are sold either as spare parts to repair shops, or directly to car manufacturers to be used on the assembly lines. The repair shops order small quantities when needed. The car manufacturers offer long term contracts usually at a price lower than the price that the shops pay. The local repair shops market is sensitive to price and quoted lead-time. The market is very sensitive to due date performances- any delivery that is late hurts the reputation of the company in the market and causes a decline in demand.

The simulation starts on 1.1.96. The company suffered losses during 1995 (about $68,500) but its due date performances were excellent (DDP=100) and sales total about 973,000. The owners decided to hire new management (you) in an effort to improve performances. Starting on the 1.1.1996 it is your responsibility to run the company and to make a profit while keeping due date performances as high as possible (at least at 80%).

The shop floor

Information about the manufacturing facility, the product structure, production times and setup times is available in the production view under information. The following points are important to remember when running the factory: The plant operates five days a week, eight hours each day. In this scenario it is impossible to operate a second shift or to work over time. There are four work centers on the shop floor: GT, M1, M2, and PK. In each work center a dedicated operator operates a single machine. The machines are subject to random failures (breaks). The three products follow the same routing (a flow shop) as can be seen in the all-product routing screen under information in the production view. Four raw materials are used for the manufacturing process. These raw materials can be purchased from two competing suppliers as explained later.

Marketing

The three products A1, B1, and C1 are sold in two markets:

1. Local repair shops using A1, B1, and C1 as spare parts- this market is very sensitive to the quoted price and the quoted lead time - the marketing policy adopted by management. A lower price and a shorter promised lead-time increase demand. The market is also sensitive to the quality of the order fulfillment process measured in terms of deviations of actual delivery time from the promised delivery time. Deterioration in Due Date Performances (DDP) damages the demand in this market significantly.
2. The contracts market, which is based on Request for Proposals (RFPs), issued by car manufacturers. Each RFP specifies the required quantities, the required delivery time and the price the customer is willing to pay. Management has to decide whether to take a proposed contract or to reject it. Rejecting a contract does not effect future demand and future RFPs. This market is sensitive to due date performances and late deliveries might cause cancellation of a contract.

The link between marketing and the production functions is the Master Production Schedule or the (MPS). The simulation starts with several orders in the MPS. New orders from the local distributors are generated randomly and can be forecast by the system. Potential contracts or RFPs are presented to the managers from time to time. Marketing has the authority to override the ERP system and to issue manual work orders.

Purchasing

The four materials used in the production process can be purchased from two different vendors. The cost per unit of each material as well as the shipping cost per shipment is presented in the purchasing view. The current purchasing policy is based on the (s, S) policy- whenever the current inventory level drops below s an order is issued to the current vendor. The order size is the difference between S and the current inventory level. A control system is available that alerts management whenever the inventory levels is below a predetermined "red level ".
The information available in the purchasing view includes past demand of each material and a forecast of future demand based on exponential smoothing.

The Finance view

This view supplies information related to the finance and accounting system. The main screen in this view is a Profit and Loss Statement that shows a loss at the end of 1995. Information on the marginal profit of each product and each contract is also available. At the beginning of 1996 the cash position is about $42,000 and a credit line of $75,000 is available from the bank. If the credit position reaches a value below the credit line, the simulation terminates due to bankruptcy.

The simulation period of this scenario is six months and the user has a total of two hours to run the six months period.

The user should review the policy used by the previous management and analyze the relationship between past policies and the value of performance measures recorded last year. New policies should be developed and implemented based on the goal of the organization along with the appropriate control system parameters.

A.2.2 The job shop

This scenario is identical to the flow shop except for the routing of the three products - each product has its unique routing on the four work centers.
The main question in this scenario is whether the policies developed for the flow shop are adequate for the job shop and if not how should these policies be modified.

A.2.3 Assembly

This third scenario depicts a new product, which is assembled from the three components A1, B1, and C1 that were introduced in the flow shop and the job shop scenarios. In this case the problem of coordination between the production plans of parts required for the same end product should be addressed. Another issue is the flexibility to use A1, B1, and C1 either as parts or as components for the end product.

INDEX

License for the Operation Trainer

MBE Simulations Ltd. grants the purchaser of the book: "Enterprise Resource Management: The Dynamics of Operations Management", by Prof. A. Shtub, the license to **one installation** of the Operations Trainer on a PC computer or a notebook. It is not allowed to install the Operations Trainer on a network.

The use of the Operations Trainer is limited to personal learning only. It is strictly forbidden to use the software for any commercial use. It is not allowed to use the Operation Trainer in class – unless the university purchases a university license from MBE Simulations Ltd.

All copyrights in the Operations Trainer shall at all times remain the property of MBE Simulations Ltd. The user is not allowed to introduce any modifications to the software and to the label on the diskette.

The user may not reverse engineer, decompile, list or print the software.

MBE Simulations Ltd. does not and cannot warrant the performance of or the results that may be obtained by using the software. MBE Simulations Ltd. hereby specifically disclaims any and all express and implied warranties with respect to the software.

Under no circumstances shall MBE Simulations be liable to the user or any other person for any special, incidental or consequential damages, including, without limitation, lost profit or lost data, loss of other programs, or otherwise, and whether arising out of breach of warranty, breach of contract, tort (including negligence) or otherwise, even if advised of such damage or if such damage could have been reasonably foreseen, except only in case of personal injury where applicable law requires such liability.

Installation of the Operations Trainer

For Windows 95:

1. Insert the diskette into the floppy drive.
2. Click on "Start", then click on "Run"
3. Type "a:\setup" and click on OK. If the floppy drive is B – type "b:\setup".
4. Follow the instructions.

For Windows 3.1:

1. Insert the diskette into the floppy drive.
2. Click on "Files", then click on the "Run" entry.
3. Type "a:\setup" and click on OK. If the floppy drive is B – type "b:\setup".
4. Follow the instructions.

The Operations Trainer and the full version of MICSS – Management Interactive Case Study Simulator

The Operations Trainer's scenarios and features were carefully chosen to satisfy the needs of this book. However, the Operations Trainer is just a limited version of a management education tool called MICSS, Management Interactive Case Study Simulation. The main objective of the full version is to provide a learning platform for real world managers through workshops and self learning.

With its additional capabilities MICSS is designed for a variety of workshops. The topics include *Management in the ERP Era*, *Integrated Resource Management*, *Inventory Management*, *Production Activity Control* and *The Theory of Constraints Challenge*. The length of a typical workshop is between two and three days. In such a workshop the participants initially face a tough case study to run for a year (simulation time). Typically the first run is not too successful, which opens the mind to new ideas about the system, the policies, the processes and the integrative thinking that is needed so much. With the aid of the instructor, the participant learns to verbalize the cause and effect in the virtual environment. This understanding leads the way to come up with better policies, focused information needs and a superior control system to guide management to make the best of an uncertain environment.

The full version of MICSS contains many additional features, including a wide variety of different scenarios with a time frame of a whole year: more products, more machines, more materials, complex routings, seasonal market demand, various graphs, option to buy more machines, going to second shift, advertising, detailed MRP module in Production and in Purchasing and outsourcing of components.